Reasons to Rejoice

Two Friends Look Back on a Life of Miracles

To Rebecca
Please Enjoy!!!

Steve Rydell
and
Bobby Schmittou

ISBN 978-1-64492-303-0 (paperback)
ISBN 978-1-64492-304-7 (digital)

Christian Faith Publishing, Inc.
832 Park Avenue
Meadville, PA 16335
www.christianfaithpublishing.com

Some of the names and locations are fictitious.

Printed in the United States of America

To my mother and to the memory of Randy Pemberton's mother and Bobby Schmittou's mother.

Contents

My Prayer

My heavenly Father,
I love you, I need you, I care about you, and I
support you in your struggle against evil.
Thank you for my many blessings. Please continue
to bless me so that I may bless others.
Thank you for Jesus, who gave his very life for me.
Help our nations in their fight against evil.
Bless those people who are unable to care for themselves.
Please have the Holy Spirit intercede and pray on my behalf.
In Jesus' holy name,
Amen

Chapter 1
Orphanage Days

Cloudy, breezy days always remind me of the orphanage.

The upside of living in an orphanage is having 150 friends to play with at any given time. The downside is knowing that since your parents are the managers of the place, you may not see them that often. Located just outside of Richmond, Kentucky, we lived in the home from the time I was eight years old until I left for college at eighteen.

Life at the orphanage was never dull. We all had our chores. The girls were responsible for the cafeteria work, laundry, and cleaning their dorms. The boys were responsible for farm work—feeding the cattle, hauling hay, maintaining the vegetable gardens, slopping the hogs—plus keeping their dorms clean.

The home was a Church of Christ-supported organization. As kids, we all attended church services every Sunday—morning and evening—and every Wednesday night. If there was a Church of Christ revival meeting anywhere within thirty miles of the home, all the kids were required to attend. We were trained to believe that only Church of Christ Christians would be saved and go to heaven. To say we were brainwashed might be an understatement.

Since my parents were the managers, my siblings and I were often treated like all the other orphan kids. My mother and father were hardworking people with big hearts for the serving in the church, which I had thought was why they signed up to be the parents of the home. My mother, Sharon McKinley, grew up in a close-knit family in Southern Indiana. As one of six kids, she had to learn to fend for herself early on. She became an excellent baker and avid reader and taught herself how to play the piano. She always enjoyed being around kids too. Her family always attended Church of Christ churches, and she followed suit as she became an adult. Her father, my grandfather, ran a hardware store, and I can still recall the smell of that store to this day. I loved it. All the tools, the concrete, the steel, the old wood, it all had this old-fashioned, musty smell that I loved being around.

My father grew up on a farm during the Great Depression, and it made a lasting impression on him. His mother passed away when he was six years old, so his father raised him, which, in those days, meant he had to work all the time. Living on a farm, there is always work to do. My dad learned to appreciate the value of a dollar and little patience for idle time. He liked sports as much as the next guy but rarely had time to play. He was all business. Like many other people of his generation, he knew how to stretch a dollar and would save as much as he could. My father also had a heart for the church and decided to attend Lipscomb University, a Church of Christ college. He then went on to get a theological degree from Peabody College of Education and Human Development at Vanderbilt University. He soon became a preacher in the Church of Christ, often filling in for various churches around Kentucky and Tennessee.

He and my mom decided early on they would take on the orphan's home as a form of ministry, but after a while it became clear my father was more comfortable either behind the pulpit or working with his hands. He would occasionally plan a big getaway for the kids, like taking a bus load of kids to see the Cincinnati Reds play a game at home or taking everyone to a summer camp in Tennessee. He loved the kids but had a hard time connecting with them one on one.

My mom on the other hand had a real knack for working with kids. Every year she would host a big tent meeting that served as a combination of a celebration, a fundraiser, and an adoption drive for some of the orphans.

Mother would kick into high gear when it came time to plan for the event and was a real professional when it came to hospitality for the guests. She wanted the best for the children. People would come from miles around for the week-long celebration, bringing home-cooked meals and, in some cases, sizable donations for the home.

But there were also hard times at the home. Whenever there was trouble, my father was the one to instill the discipline, and a few times he was pretty hard on a couple of us. When I was nine, my middle brother, Ray, had a bad habit of picking on my younger brother, Joe. When I would catch Ray in the act, I would pop him on his shoulder, and Ray would run crying to Momma. When my father got home that night, he punished me mercilessly. I won't go into details, but over the years I began to experience mental and emotional health problems. Some psychologists say living in an orphanage for more than five years can be psychologically harmful for any child. For the most part I had fun with my brothers and friends while living there, but there was something else going on inside of me I couldn't quite explain. I would later be diagnosed as manic depressive.

Barnyard Antics

Living out in the country with animals does provide some interesting situations. When I was eleven years old and Joe was four, we were playing in the barnyard behind our house. Joe bent over to pick up a rock, when suddenly, a large rooster attacked him. Thinking on my feet, I noticed an old rusty basketball rim laying on the ground and swung into action. I grabbed the rim and slung it as hard as I could toward the big bird. The rim swished through the air for what seemed like an eternity before finally striking the rooster and ringing its neck. The bird flopped on the ground for several minutes before eventually expiring. My brother was safe, and that was all that mattered to me.

Now I was only eleven years old at the time, but I can't help but look back and wonder if there was angelic intervention. What would have happened if I had hit Joe instead of the rooster? Angel interference? I sure hope so. In fact, I've come to believe it was just that.

I've had my share of bad luck as well. During my early years in elementary school we lived in a small town in Tennessee called Ripley. That's where I met the McCutcheon twins, Ronnie and

Donnie. One day, back in the first grade, we were climbing trees at their home, having a great time just being boys. I eventually found myself about fifteen feet up the tree, and without realizing it got myself out on a dead branch. The branch snapped, and I went flying. Falling headfirst, instinct told me to stick out my arm to break my fall. A broken elbow and a significant amount of crying later, I was rushed to the hospital and underwent surgery to reset my elbow. To this day I still have the six-inch scar it left on my right arm. As bad as this was, I still consider myself blessed to not have become paralyzed.

Back in Kentucky, my parents eventually accumulated enough extra money to acquire a small forty-acre farm on the outskirts of Richmond. It had a big barn and a small tenant house with dog pen. They ran about fifteen head of Herefords and steers on the acreage, as well as a vegetable garden.

One cloudy day, when I was about twelve, my dad and I were moving the cattle from one field to another, when "Old Crazy," a big steer standing about four-and-a-half feet tall, wandered off and managed to get into the dog pen. We followed him, when suddenly, the steer turned and attacked my dad, knocking him to the ground, and butting Daddy's head on a rock.

Without a moment's hesitation, I took off running from ten feet away and tackled Old Crazy. He didn't move, probably being in a state of mild shock. My dad then yelled, "Come on, Steve, let's get out of here." We left the steer in the dog pen and stopped for the day.

I attribute my "superpower" to either an adrenaline rush or a second episode of angelic assistance. After all, tackling a four hundred-pound beast doesn't happen every day. At the very least, my actions showed courage and faith and even my devotion to my father.

Horses

In the summer of 1969, the year of the lunar landing, I received a birthday gift I'll never forget—a pony. He was beautiful, with brown and white spots, and the perfect size for me. I can't remember where my parents had bought him, but his given name was Lightning.

My first day's experience with Lightning was memorable as well. With the help of the farm manager, we put on his new saddle and bridle. Then, upon mounting him, he reared straight up, knocked me off, and fell back right on top of me. Although I was unhurt, except for my pride, it was later I realized Lightning quite possibly had never been broken. Maybe he had simply reacted as he would if attacked by a wild animal.

Oh, he had more tricks in store for later days. He loved to ride along the fences and try to rub my leg up against them. Also, he liked to be running at full speed across the field, and then just suddenly stop, with me trying to hang on for dear life (sometimes successfully and sometimes not). I guess my favorite trick of his was when he ran right through the limbs of a tree that had recently been struck by lightning. He managed to scratch up my face bad.

Over the next few days, he settled down, and there were fewer incidents. I loved Lightning and rode him virtually every day over the next several years. He was the first animal I truly loved and enjoyed taking care of. I learned from Lightning that even though a relationship may not begin with "love at first sight," it certainly can develop over the long haul.

As I grew older, I graduated up to the full-size horses at the home.

And then, the McCutcheon twins came to visit us in Kentucky. I should have known I would be in for the ride of my life that day. By then, I had a favorite horse named Ladybird, after the First Lady. She was indeed a "black beauty." Somebody had already readied Lady Bird to ride when we arrived at the barn. I mounted Lady Bird, and we took off at a snail's pace toward the woods.

Usually, we would ride our horses at a walking pace to the woods about two hundred yards away from the barn. Then, we would turn the horse back toward the barn, and they would gallop back. This time, though, upon making the turn to go back to the barn, she suddenly went crazy. She was running, bucking, and rearing up, at breakneck speed. Finding myself in the middle of this unplanned rodeo, I was thrown right over her head, landing on my head. Miraculously, I didn't break my neck, so I ran home yelling, "Momma, Momma." They immediately took me to the hospital, where I was diagnosed with a severe concussion.

Later I was informed that Lady Bird's saddle had been incorrectly strapped around her flank area, causing her to react the way she did.

It appears the McCutcheon twins had brought me bad luck once again.

Randy Pemberton

We had many friends at the home, but none were as good looking as Randy Pemberton. He was so handsome we gave him the nickname Lover Boy. But his personal story, like so many other orphans at the home, was rough.

Randy began fending for himself at six years old when his mother dropped him off in the projects in Nashville. After a couple of years living on the streets, he was finally discovered and sent to a Church of Christ institution in Western Kentucky called Paradise Home. Needless to say, Randy was unhappy there and often ran away. As a result, he found out the hard way how they treated runaways. The managers at Paradise treated them like dogs, tying them to a tree for three days, setting a bowl of food and water in front of them.

Later, Randy was sent to our orphanage outside of Richmond, and he quickly earned the nickname Lover Boy. He was also a very good athlete and made the high school basketball team. Randy became a superstar, with his outside southpaw jumper.

Randy had the best sense of humor, even with the hardships he endured. He had a joke a minute. Seemed like he was always ready with a funny comment. One of my favorites was, "When things get rough for most people, they get just right for me."

Still unhappy with the orphan's home environment, he one day decided to walk back to Nashville. Yes, walk. He set out early on a Sunday, following the railroad tracks for much of the journey. When he reached Franklin, Kentucky, however, he went right by the Franklin Church of Christ just as Brother Rydell—my father—was walking out, having just preached there.

Seeing Randy, he yelled out, "It's about time for supper, isn't it?"

Randy agreed and rode back to the home with my father. Exhausted from his journey, Randy was put to bed early that night. We were all glad to see him again.

A few years later Randy was taken in by foster parents in Hopkinsville, Kentucky. Mr. Eastly was a song leader in an area Church of Christ. They treated him like a slave. He was afraid to even ask for a Coke while working on their big farm, for fear that they would send him back to the home. For all his hard work, they promised to put him in their will but reneged on that.

Later in life, Randy would build his home with his own two hands, having bought eighteen acres for $1,500. Randy became a successful self-employed stone mason, with his work being in demand for miles around in Tennessee and Kentucky.

The Sporting Life

I and my friends loved to play sports. Any kind of sports—baseball, softball, basketball, football—you name it, we played it. As far back as I can remember, we had pickup games as often as we could. We all attended grade school at the orphanage and would later attend the county high school on the other side of town.

Being of medium build and height, I never had much of a chance at playing organized basketball in high school. However, one day we had a much-anticipated pickup game at the home involving our best players against some of the guys from the high school. We planned to play the game in our new orphanage gymnasium, complete with referees, scoreboards, fans in attendance, and all.

My best friend, Terry Widner, and I were average basketball players. Terry, though, was what I like to call a creative thinker. Prior to the game, he suggested we drink a bottle of Boone's Farm strawberry wine he had hidden in his room. Sounded good to me. We grabbed the bottle and headed out to a wooded area on the farm. It would be my official introduction to alcohol, so Terry gave me the first taste. It was so sweet I drank half the bottle before handing it back over to Terry. He drank the rest, and we made our way over to the gymnasium.

The game was hotly contested at first, and then Terry and I seemed to catch fire. Thanks to the wine, my confidence spilled over into overconfidence, and I thought there wasn't anything I couldn't do on the court. By the second half, I kept telling Terry to pass me the ball, yelling, "I'm open, I'm open!"

We had a blast and ended up winning the game by eighteen points; I ended up with thirty-five points, and he had twenty-six points.

When the euphoria of the game wore off, I began to feel sick to my stomach. But I didn't care because we had so much fun. Being my first experience drinking alcohol—and the last for a long time—I learned it is okay to drink in moderation if you're not putting yourself or anyone else in danger, such as driving. Of course, being raised as in the Church of Christ, any drinking at all would have been frowned upon.

I had a lot of good friends in those days. But I wasn't always careful with my mouth, and it cost me one day. Sitting in the high school cafeteria eating lunch one day with some of my friends, our popular black friend, Albert Cobbe, joined us. We loved hanging out with Albert. While we were shooting the bull, one of the guys made a comment that there was no difference between blacks and whites. To which I made a quick, stupid, sarcastic remark that did not go over well at all. A bad teenage attempt at humor. It obviously hurt Albert's feelings and he immediately left the table. I still regret that comment to this day. So, if you are out there Albert, please forgive me. I love you, brother!

Bullies and Football

Even though I didn't have a large build as a kid, I still took it upon myself to defend my friends against bullies. Some victims included my first-grade friend, Barry Harrel, Brent Holton, and even my brother, Joe. Brent was kind of gangly and awkward, not very athletic. This evidently made him an easy target for bullies. Even though he stood 6'2", he found himself a target of our freshmen football team.

When I caught them making fun of Brent, I would jump in and warn them to back off. Thankfully they did. Turns out the players were simply mimicking their coach, who turned out to be the biggest bully of them all. Coach Burleson was an old-fashioned, tough-nosed man who had little sympathy for weak men.

I decided to play football that year and had no choice but to play for coach Burleson. Being shorthanded as a football team, some of us had to play on both offense and defense. So, I played wide receiver and defensive end. We had a good team. We were undefeated going into our final game against our crosstown rival.

On one particular play late in the game, an enormous guard was coming right at me. My job as a defensive end was to keep him and the play to the inside of the field. This guy had to be 6'5" and 350 pounds. I didn't have a chance. He easily bowled over me, crushing my right knee. A couple of our players helped me off the field. We went on to lose the game, and there went our undefeated season.

After riding the bus back to our school, we all headed to the locker room. Coach Burleson, a big man at 6'1" and 300 pounds, approached me while I was sitting on a bench. He lifted me up by my jersey and told me that he knew I was faking this injury.

Later, when I had knee surgery, he came by the house while I was recuperating. He did his best to apologize by bringing some of his wife's homemade banana nut bread.

At the end of the school year, we always had an awards day. Coach Burleson called me over to his office on the loudspeaker. As I entered his office in the gym, sitting there were both Burleson and assistant Coach Morris. They informed me that although I hadn't finished the season, they would award me a football letter, on the condition that I come back and play next year. It only came to my attention years later that Coach Burleson was a member of the Church of Christ.

My love for sports would nevertheless take me in a different direction.

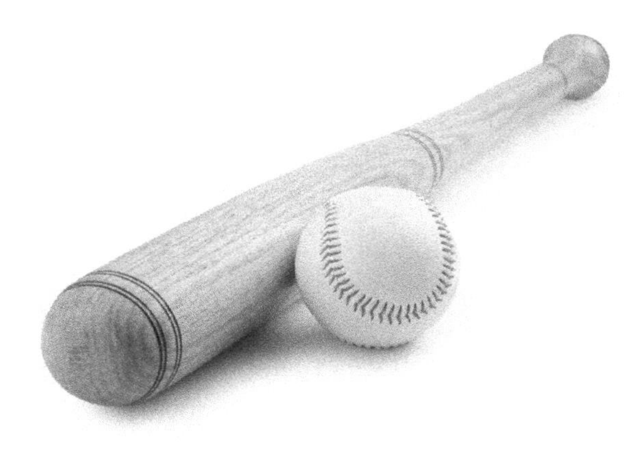

Chapter 2
Baseball and a Family of My Own

After getting injured playing high school football, I turned my sights to baseball and found my true love in sports.

My professional baseball hero at the time was Pete Rose of the Cincinnati Reds, also known as Charlie Hustle. I just loved the way he played the game all-out. I played second base and did my best to emulate Charlie Hustle. I would slide into every base head first, including home plate. (I still hope Pete makes it into the Hall of Fame. I am one of those fans who have forgiven him for off-the-field antics.)

Being slim, my asset was my speed. On every team I played, I was given the green light to steal bases at any time. I maintained a good batting average too, hovering between .300 and .600 during my high school career. I worked hard to hit a lot of doubles and triples.

During my senior year, I had four game-winning hits.

One beautiful spring day, we had a doubleheader in Bowling Green, against Russellville High. I felt great—my high school sweetheart, whom I would later marry, Misty, was going to be at the game. Going to my favorite sporting goods store, I purchased a Pete Rose 34' wooden bat. The rest is history.

I went eight-for-eight that day, with a hit in every at bat. Not being a power hitter, I even hit a ball that bounced off the top of the center field fence. While rounding third base—thinking I had an easy inside-the-park home run—I was humbled. Suddenly I stumbled and hit the dirt. It was all I could do to scramble back safely to third base. Oh, my teammates had a blast with that one. There were all kinds of cheering and commotion. Thankfully, we won both games easily, and I believe Misty was impressed.

Later that same year, we made a road trip to Gamaliel High. During the ninth inning with a runner in scoring position to tie the game, Coach Jones gave me the signal for a sacrifice bunt. Laying down a bunt that rolled slowly toward third, our runner scored easily.

I'll never forget on the ride home on the bus, Coach Jones stood up and proclaimed, "Rydell, you played a helluva game!"

Another memorable quote from him came during our tryouts earlier that season. Being the only senior in a group of mostly juniors, I was taking batting practice. Robin Popeye was a junior who was trying out for second base like myself. One of the juniors yelled out. "Rydell, you'll never make this team!"

To which Coach Jones yelled out. "He will if he keeps on hitting that way!"

Once, during a summer league, I faced a pitcher named big Gordon Jackson. He stood about six feet six inches and was also a star basketball player at Bowling Green High. Well, I heard that he had a crush on Misty. The first pitch from Gordon was a screaming high fastball aimed right at my head. That ball nailed me in the neck, right under the helmet. It knocked me flat on my ass and hurt like hell. But it didn't knock me unconscious. Although I was mad, I kept my cool, got up and dusted myself off, and trotted down to first base.

The game was on. On his next pitch, I stole second base. Then, on his next pitch, I stole third base. And you better believe, on the very next pitch I stole home! Sweet revenge!

As the school year began winding down, I began thinking about my future. My parents, of course, were adamant about me going to a Church of Christ college. I loved playing baseball and was really hoping to land an athletic scholarship. Thankfully I had paid attention in class during high school and graduated third in a class of two hundred. So I was also eligible for an academic scholarship.

My folks pushed hard for me to go to Lipscomb University, in Nashville, Tennessee, a small Church of Christ college. I scheduled a visit with their baseball coach, Ken Dugan, about the prospects of coming down to play for Lipscomb. He basically told me that if I make my summer league all-star team, he'd come up to watch me play.

Well, I batted .458 and a slugging percentage of over .700 and led my team in stolen bases—as I had done on every baseball team I ever played on. But what I didn't know about their summer league was that the players had more sway than the coaches. And since the players picked the all-stars, and I was the newcomer, I didn't get the votes.

Although frustrated, I decided to still attend Lipscomb and tried out for the baseball team as a walk on. After all, in those days, Lipscomb was the dominant school in the NAIA (National Association of Intercollegiate Athletics). So why not try to play for the best?

Again, I was trying out against a bunch of scholarship players. Coach Dugan never got to see me play in a game situation, and I didn't make the cut. I loved playing baseball, but I knew my playing time was coming to an end. I enjoyed my years at Lipscomb. Misty and I began growing closer together.

About that time, Misty and I decided to get married. I always enjoyed being around her parents, Joe and Nancy Pippen. I first met them in my junior year of high school, while dating Misty. I consider them the first "real" people I met outside of the Church of Christ circles. There was genuine love in that family that I admired. They were Episcopalian, but they didn't take it too seriously and rarely attended church services. The Pippens taught me so many things like how to water ski. Joe taught me how to hunt squirrel and doves. In a way, they were like parents to me. They both enjoyed an occasional toddy, and Nancy even smoked cigarettes.

I learned that you don't have to be Church of Christ to be "real" people. By this I mean that with the Church of Christ, I have noticed so many members spend most of their non-worship time with church activities. It's almost a country club environment.

So, I married Misty during my senior year in college, and due to some weird rule, I lost my academic scholarship. So, I started working nights, thirty hours a week, in addition to full-time college. While working at H.G. Hill's grocery store, stocking shelves, I met

Bobby Schmittou, who to this day remains one of my best friends. Bobby and I played softball together on the company team. I still managed to graduate from Lipscomb in three years' time.

During my college years my beliefs about Christianity would change, thanks to a new friend I made at Lipscomb, Stan Level. Standing at 6'6", Stan broke almost all the basketball records at Lipscomb and would also serve as our senior class president. Stan opened my spiritual eyes when we were golfing one day. Stan flat out stated, "You know, Steve, I don't believe that Church of Christ Christians will be the only ones saved."

That hit me like a ton of bricks and changed my life forever. Since then, my spiritual path has led me to be a member in three different Christian denominations. It has been such a blessing. These various denominations held most of the attributes of the Church of Christ—worshiping on Sundays, offering communion every Sunday, and fellowship outside of church. Literally, the only difference was the use of instrumental music in worship. Because Stan loved me enough to show me the difference, I now have grown closer to God and the Holy Trinity due to these experiences.

I eventually took a job as a certified public accountant and found that I was pretty good at it. Misty and I settled down and soon had two kids, our son, Mark, and our daughter, Sherri. Life was good, but I began spending more and more time at work and not at home. I took on as many clients as I could to make as much money as possible. Soon I found myself working ninety-five hours per week for months at a time. This continued for several years in a row. Then I began experiencing terrible sleep patterns due to my profession as a CPA. Looking back, I should have simply hired more help during those periods. Hindsight is twenty-twenty.

One thing Misty and I would do to relieve stress is take vacations. We took one memorable trip to Cancun, Mexico. It started with a beautiful boat ride out to the Isla Mujeres, accompanied by some barracuda, who swam alongside the boat. Upon reaching the island, we noticed tourists renting snorkeling equipment at the shore. We chose, however, to climb a bluff to a quaint restaurant and have a few drinks and something light to eat. After three "Free

Cuba" drinks, I was feeling no pain. Off in the distance, I noticed a large rainbow-shaped cove at the end of the island. So I decided to swim over to it and search for treasure. Descending the bluff, I bypassed the vendors and entered the water. Halfway out on my little swim, I realized I may have undertaken too big a task. I had to decide whether to go forward, or return, with it never entering my mind to go to the shore. I decided to proceed.

After making it to the cove, I looked around for the treasure but found none. What I did find, though, were some pretty conch shells. I stuck two or three in my swimming trunks and carried one in each hand. Set for my return swim, I set out. By now Misty is probably wondering where I went. Or she could have been counting the half a million dollars in life insurance she would get upon my untimely demise.

The swim back was easier, for some reason. I climbed the bluff and happily showed Misty the conch shells. She smiled and loved seeing the shells. We enjoyed the relaxed atmosphere, the gorgeous views, and the time alone to decompress.

Later, we would take a trip to Florida with our close friends, the McAllisters, with very different results. Since Joe McAllister was a big Vanderbilt fan, our first stop on the trip would be a detour to Athens, Georgia, to take in the Vanderbilt versus Georgia football game. Well, on the way down from Nashville, Joe and Terri fired up some pot in their van. Misty immediately became upset and told them she didn't appreciate them smoking pot around her. Joe stated that it was their van, and they would smoke if they wanted to. When we stopped at the hotel the first night, in Marietta, Georgia, I tried to ease the situation with Misty. I reminded her that her uncle owned virtually every liquor store in her hometown, and I never heard her preaching to him. After all, liquor causes more devastation than pot.

That didn't seem to help the matter, and she continued to pout the rest of the trip, including even during the football game. After growing tired of her attitude, we stopped for the night in Montgomery, Alabama. In the hotel room, I resorted to telling her that if her attitude didn't improve by morning, I would put her on a flight back to Nashville. Morning came, she did not change a bit, and I put her on the plane.

Meanwhile, the rest of us went on to Destin and had a great time. We were mainly fishing for king mackerel on the piers, and then Joe, who was in the catering business, would grill them out. Some seriously good eating. The two negatives from the trip: I began smoking cigarettes, and the beginning of the end of my marriage was near.

I found out years later that Misty told my parents and a few others that I had participated in the pot smoking. I had not. I suspect that's why my reputation began to suffer at church. Yet I forgive them all for their gossiping. Rejoice in the Lord!

A most enjoyable trip involved myself and about a dozen other members of the Belmont Church, on Music Row in Nashville. We flew to Mexico City and then on to Lake Guerrero, on a bass fishing excursion. What a blast! Each day started with the camp guides fixing an elaborate breakfast, and then we were out on the lake by sunrise.

Each of us was catching around twenty "keepers" a day that would later serve as our supper, again, as cooked up by the camp guides. I even caught the largest bass, weighing about seven pounds. It put up quite a fight. Guerrero being a dammed-up lake, and flooding large orange orchards, one would often see the tops of the trees above the waterline.

Scott, who was teamed up with me in our boat, along with our guide, Jose, soon learned that Jose's hobby included aiming straight toward one of these treetops and then veering off at the last second.

On one such occasion, I was sitting in the front of the boat, and Jose failed to veer off and went right over the tree. He started yelling something in Spanish, which we did not understand. Evidently, we were taking on water fast. Jose, to his credit, headed toward the nearest land. And we got there just in time, as the boat started to sink. Then Jose ran off into the thick brush, leaving us stranded. Eventually, we were able to flag down a passing boat, with a couple of guys from Houston. They picked us up and carried us back to our camp. That was the best $700 I've ever spent on a trip.

It was Halloween, and when I returned home, my family greeted me at the airport. Mark, about four years old at the time, was dressed like a cowboy and stated that I should grow a mustache like him. He evidently didn't notice that I had started my first mustache while on the trip. I've never shaved it since.

Speaking of my son, Mark, on two occasions I helped save his life prior to his reaching nine months old. On the first such occasion, Misty had left me with Mark, while she went shopping. She gave me no instructions on how to feed him, just a filled-up bottle with formula. When he started to cry, I figured it was time to feed him. I did the best I could with him and the bottle, but with no practice, I felt a little lost. And, as fate would have it, he began chocking. Moment by moment, it got worse, and I didn't know how to stop it or help him. Soon, his eyes rolled back, and he became unconscious. I quickly dialed 911. I then turned him upside down and patted him on his back. Thankfully, this revived him.

When the ambulance arrived, I asked them to play it safe and go ahead and take him to the hospital. Following in my car, they later told me that Mark would be fine. Later, upon reexamining that bottle, I discovered that the hole in the nipple was too large, and the milk flowed out freely just by turning the bottle upside down. He could have drowned. I thank God for his intervention that day.

The second incident occurred when Mark was nine months old. He had been sick for several days, with a fever, and while Misty was a stay-at-home mom at the time, she didn't notice that his belly had swollen to the size of a bowling ball. Upon getting home from work on Friday, I did see the difference in his belly and said we'd better get him to the doctor immediately. We drove quickly, covering the distance from Tusculum to Green Hills in fifteen minutes. After a quick examination by Doctor Mallard, he stated that we had better get Mark to Vanderbilt Hospital immediately. We drove quickly to Vanderbilt, and Dr. Mallard called ahead to prep for emergency surgery.

It appears that Mark had major blockage where his large intestines meet the small intestines. They rushed him into surgery. Although the surgery was a success, the surgeons informed us that if he had gone another couple of hours, we would have lost him. Once again, the hand of God was with us.

I truly believe these incidents created a special bond between me and Mark, much stronger than the typical father/son bond.

A Highway Miracle

By the time I reached my early thirties, I was officially diagnosed with manic depression. The doctors were trying different medications to suit my needs and analyzing my odd sleep patterns. Well, something wasn't working right, because on one fine day while driving down the highway, I decided to end it all. I didn't have any big reason, but it was obvious to me the doctors had not yet perfected my medication.

As I removed my seat belt, I drove off the road at sixty miles per hour, took out a utility pole, snapping it like a twig, and then barreled down a steep embankment. My SUV should have rolled by now, but it did not. Then I slammed straight into a large tree, which stopped me in my tracks. As I assessed the scene, I had not even so much as a bloody nose.

I climbed back up the hill only to find a man sitting in a black pickup truck, waiting for me.

"Do you need a ride to your doctor?" he asked, in a baritone voice.

"Yes," I said. "Thank you."

Not a word was spoken as we drove to my psychologist's office. I never even gave him directions. Finally, as we arrived at the office, the gentleman stated, "Never try this again, because we have big plans for you in the Kingdom's work."

A miracle to still be alive? I know so.

An angel of the Lord in my presence? I believe so. Rejoice in the Lord!

Chapter 3
Miracles, Laughter, and Risky Business

After my marriage to Misty ended, I spent increasingly time on my business life, which, as it turns out, seemed to get more interesting by the day.

I had invested in a start-up company which needed to raise capital of about five million dollars. The theme of the company was the old west and was to be based in Wyoming. My partner, Lewis, had controlling ownership. Lewis was a nice enough guy, with a winning, charismatic personality, but I did not sense he was a Christian. On one occasion, I lined up a meeting with the largest venture capital firm in the south, based in Nashville, where both Lewis and I lived. They agreed to fund our organization, but because we did not react quick enough to get our attorneys involved, the venture capitalists changed their minds and backed out. About that time, unbeknownst to me, Lewis perceived me to be a threat to his control of the company.

By this time, I wanted to see our venture in person in Wyoming, so I planned a trip during the summer. We would drive out West to Yellowstone Park in Wyoming. Mark and Sherri were fourteen and eleven at the time and ecstatic about the trip.

We set out in my Mitsubishi 3000GT sports car, which got reasonably good gas mileage but had superb power if you ever needed it. We took our time, making about five hundred miles a day and stopping at Best Western's at night.

We toured Mount Rushmore along the way. The kids were impressed with the size of the famed statue, as well as the detail of the facades of Washington, Lincoln, Jefferson, and Theodore Roosevelt.

Then we set out again for Yellowstone, with Billings, Montana, our next scheduled stop. I took a short break along the way to call Sonny, a friend of the company. And he invited us to stay overnight with him and his wife and then go by their cabin in Red Lodge on the way down to Yellowstone. Note: this was before the days of the cell phone.

Sonny's wife, Lisa, had fixed a great home-cooked supper for us by the time we got there that evening. After visiting for a couple of hours, we all turned in for the evening. I and Mark and Sherri slept on sofas in the living room. In the middle of the night, the phone rang, waking me out of my light slumber. Then things got weird. I could hear Sonny and Lisa arguing loudly from their bedroom.

That next morning, after enjoying sausage, gravy, and biscuits for breakfast, we were heading out the door for the cabin, when Lisa stated, "I wish we could have met under better circumstances." A very ominous statement. Also, evidently, overnight, the plans for the day had changed. Sonny would wait for a friend, whom I had heard of, Mick, and then they would later meet us down at the cabin. He gave us directions, and we were on our way.

We reached the cabin soon and pulled up some lawn chairs to sit and wait for the others. It was a beautiful place, sitting on about three acres. The gravel driveway came up the slight hill, then looped around, forming a circle, and then headed back downward. The drive was surrounded on virtually all sides by tall lodgepole pines, and any clearings were covered with tree stumps and large boulders. Due to the events of the night, and Lisa's statement, I was on *alert* mode.

Back in 1986, I almost went with the FBI, passing all their interviews, local exams, and the background check. I only pulled out before going to Camp Quantico when I realized that their pay scale was much smaller than I had anticipated—not enough to meet my financial obligations at the time.

My next red light came when Sonny and Mick drove up in Sonny's pickup truck, literally parking only halfway up the driveway and totally blocking my 3000GT in, because I was parked in the loop at the cabin.

After introductions to Mick were made, the kids went off to play in the forest. We grabbed a Coors and sat down in the lawn chairs, sitting close together.

Upon shooting the bull for a while, Mick asked. "You know why we're here today, don't you?"

"No," I said.

"We're going to have to get rid of you," said Mick.

"Why?" I asked, trying to be calm.

"It's over control."

"And who's behind this?" I asked.

"Lewis," he replied.

And almost as if nothing had happened, our bullshit session started right up again.

Then Mick said something funny, and I pushed him the way you do when you're joking around. Much to my surprise, he fell back right on his ass. I know I didn't push him that hard. And then when I reached to help him up, it was as if I had superman-type strength, lifting him clear off the ground. Meanwhile, upon seeing this, Sonny started backpedaling, as if he had seen a ghost.

I was halfway expecting Mick to come out with a gun or a knife at this point.

They went inside the cabin, saying they were going to fix a little supper. Suddenly, Sherri came bouncing out of the woods. I calmly whispered to her to go get in the car. I walked as cool as I could muster up, into the kitchen, and past Sonny and Mick, stating that I was going to get some more smokes out of the car. Upon reaching the car, I fumbled around like I was looking for some cigarettes and then jumped in with Sherri. I gunned it toward an area that had no lodgepole pines. It still had boulders and stumps about one to two feet high. And I'm in a car with a clearance of about five inches.

When we reached the boulders and stumps, the 3000GT literally took flight, landing on the driveway just beyond the pickup truck! Rejoice in the Lord!

But I still had to get Mark out of there. I sped toward the nearest home with a telephone pole and rushed in to ask the gentleman and his wife if I could use their phone to dial 911. They said sure, and I reported the address and what had happened. Then I asked the couple if they would mind watching my daughter a few minutes, and they agreed.

When I asked if I could borrow a gun, the gentleman said, "Mister, I don't even know you."

To which I replied, "That's okay, I've got my old Pete Rose bat."

After giving Sherri a peck on the forehead, I left, speeding back in the direction I had come.

When I arrived, four County Mounties were already on the scene, with lights flashing. About that time, I saw Mark walking out of the woods.

"Get in the car, Mark," I yelled.

Then one of the sheriff's deputies approached me, wanting my version of what had occurred. After telling him, he stated that they wouldn't be able to do anything about it since nobody was hurt.

At that, I decided to skip Yellowstone and head home. But our little adventure wasn't over yet. After picking up Sherri, and thanking the folks, we were on our way back to Billings, when I noticed off to the side of the road that a group of five or six guys scrambled to get into a red pickup truck and chase after us. This must have been their backup plan.

I kicked the 3000GT in gear once again, and it performed to perfection. I lost the truck easily on the country road, telling Mark to get out the map and find me the quickest way across Wyoming to Sheridan. I knew we would need gas soon. Meanwhile, Mark had found an ideal way across Wyoming, and we set out on our journey home, stopping only for gas and food until we reached Kansas City.

Upon our return to Nashville, I immediately called Lewis and confronted him with what had happened. Of course, he denied knowing about it, or his involvement, etc. I told him I was there, and I knew what was said. I offered to fight him, Mick, and Sonny, hand-to-hand, anytime, anywhere, and that he could go last, because I'd be a little winded by then. He was surprised at my proposal and declined the invitation. Good thing for him. Rejoice in the Lord!

The Security Profession

After returning from the Wyoming "vacation," I sought to surround myself with people in the security profession. One of my first gigs was at a nightclub located off Lafayette Street near downtown Nashville. Working as a bouncer, I was permitted to carry a nightstick.

The hot times were every Friday and Saturday night, averaging about one hundred patrons each evening. They had a large dance floor, a bar, and a live band. One main rule was no beer allowed out on the dance floor.

One night, three male patrons were dancing out on the dance floor, while drinking their beer. I went into action and grabbed each of them, turning them around and pushing them through the crowd toward the entrance, and reaching it, proceeded to throw them out. My boss—we called him Pappy—pointed to a guy sitting nearby.

"See that guy, Esteban?" said Pappy. "He's the leader of their gang, and if we don't let them back in, they'll all be waiting for you when we close down tonight."

Like any reasonable person, I responded that he might have a good point. So, we let them back in, and the night ended peacefully.

One other night, however, would not end so peacefully. There was a riot in the street after closing, with about eighty or so patrons fighting in the street. There were only four of us in security to break it up. Our boss, Pappy, was the only person who carried a gun.

When I turned to my right, I saw four tough guys marching shoulder to shoulder, entering the fray. I did my best to block them, spreading my arms out wide, and managed to back them off.

Then I spotted a guy at the top of the hill, waving his belt—complete with a six-inch buckle—as if daring anyone to approach him. Well, I did, with my hand held out as if to say, "Hand it over." He ran and jumped in a waiting car. Before they drove off, I told them to be careful. They vacated Lafayette slowly.

As I turned back around, I caught a guy driving his red pickup right toward the crowd. I started to attack the windshield with my nightstick, when the driver veered directly toward me. I had to dive and hit the asphalt as the truck sped by me. The driver then proceeded to run right over a patron with both axles. By then, police cars and ambulances began showing up. I told them to rush the victim to Vanderbilt Hospital. I was so concerned about that guy's health. I called Vanderbilt the next day, and unbelievably he had already been released. Rejoice in the Lord!

Pappy said I had performed so well; he offered me a chance to fill out the police report, which I did. But a few days later I realized how dangerous this business is, and with two young children at home, I decided to move on.

One of my security assignments was with a company called Klasko. I had the third shift, and during this shift, the entire plant and office was shut down. Part of the job involved checking in as to your location and time, by inserting a key into a key box. The key dangled from the key box by a chain. This served to indicate your location at a specific time; in other words, you were making your rounds as they were assigned to you.

At one point, I unlocked the office door and immediately inserted the key into the key box. Then, I proceeded down a narrow hallway about twenty feet long and made a left turn to go toward the next key box. Upon entering the key into that key box, I turned back to go toward the front of the office. As I walked past the narrow hallway, I heard a "*ping*" and looked down the hallway toward the doorway I had originally entered, and I observed the key/key chain spinning rapidly around like a fan.

Then, as I completed my designated rounds through the rest of the office, I passed a metal coat rack. It made a tingling sound, as if the hangers were bumping against each other, as if the aerodynamics caused by my walking by was causing it. On my next round, when going through the office, I made notice of the coat rack when walking by it. Again, there was the tingling sound, but I noticed the hangers were not moving at all.

Ghosts from the war between the states? Who knows, but I didn't stick around too much longer to learn of more of their tricks!

Real Estate

While working in the security field at night, I thought I would try my hand at real estate during the day. My interest was in commercial realty, and I specialized in large tracts of land. A broker I knew, Latham Winchester, approached me about placing my license with him because he had several buyers with extremely deep pockets. Latham was an elder with one of the Nashville area Churches of Christ. I decided to go with them.

Immediately, I got to work on "bird-dogging" properties, and I came up with three very quickly:

- 238 acres for sale in Leiper's Fork for $6,000,000
- 600 acres of development property for $21,000,000
- A thirty-story-tall office building in downtown Nashville for $100,000,000

I thought I was killing it, but when push came to shove Latham, and his buyers failed to secure any of these deals. Empty promises. Although this created some bad blood between us, I have forgiven him in my heart for having outright lied to me.

One of my best friends, known by his nickname of Valter Hairless (yes, he was bald), worked together with me and John Arno, in commercial real estate. He served as a merchant marine in World War II. Having a sense of humor that wouldn't quit, he would say just about anything to get a laugh.

One of Valter's favorite stories was called "An Atheist in the Woods." I love it so much I will retell it here.

An atheist was walking through the woods. "What Majestic trees! What powerful rivers! What beautiful animals!" he said to himself.

As he was walking alongside the river, he heard a rustling in the bushes behind him. He turned to look. He saw a seven-foot grizzly bear charge toward him. He ran as fast as he could up the path. He looked over his shoulder and saw that the bear was closing in on him. He looked over his shoulder again, and the bear was even closer. He tripped and fell on the ground. He rolled over to pick himself up but saw that the bear was right on top of him, reaching for him with his left paw and raising his right paw to strike him.

At that instant, the atheist cried out. "Oh, my God!"

Time stopped.

The bear froze.

The forest was silent.

As a bright light shone upon the man, a voice came out of the sky. "You deny my existence for all these years, teach others I don't exist, and even credit creation to cosmic accident. Do you expect me to help you out of this predicament? Am I to count you as a believer?"

Then the atheist looked directly into the light. "It would be hypocritical of me to suddenly ask you to treat me as a Christian now, but perhaps you could make the bear a Christian?"

"Very well," said the voice.

The light went out. The sounds of the forest resumed. And the bear dropped his right paw, brought both paws together, bowed his head, and spoke. "Lord, bless this food, which I am about to receive from thy bounty through Christ our Lord. Amen."

Valter always had us laughing.

Our biggest deal included the Leiper's Fork property I had my eye on. We were attempting to get the listing from the owners who lived in Green Hills in Nashville. Upon arriving at their address, and noticing their beautiful million-dollar home, and professionally manicured lawn, we went to the front door and arrived for our ten o'clock meeting time.

We were greeted by the owner's wife and her daughter, ages about sixty-five and thirty-five, respectively. The Mrs. was dressed casually, wearing a night gown, and it appeared she had rushed to put on her makeup. We went to the dining room after the introductions, and she graciously poured us cups of coffee. She then explained why her husband couldn't be at the meeting. We continued nonetheless.

John, being our broker, and the most experienced realtor, did most of the talking, and things were going quite well. At one point the Mrs. mentioned that her husband had played football for the University of Tennessee. We immediately high-fived across the table.

About that time, Valter had noticed that she had put on her lip liner in a way that it was about one-fourth inch off her lips. Much to mine and John's surprise, Valter then stated, "Ma'am, you sure do wear your lipstick in a peculiar way, but I like it."

You talk about a jaw-dropping experience. Well, after a few niceties, we were escorted to the front door. At the landing, Valter tripped, and I literally had to catch him to keep him from falling all the way to the floor. Now we looked like the Three Stooges, too.

Getting into John's SUV, he looked back at Valter and said, "If you were thirty years younger, I'd smack the hell out of you right now."

We lost the deal. But the humor of it all was priceless!

Chapter 4
Hunting with Mark and Dealing with Family

Everything I know about hunting I learned from Joe Pippen, my former father-in-law. My in-laws possessed a mostly treed farm of three hundred acres in Barren County, Kentucky. It was perfect for hunting. We had every kind of game out there, from raccoons, to squirrel, to dove, there was always something to hunt. All that time together outside also brought us closer together. He was a kind man. In fact, whenever anyone asked him what was needed in life, he would respond, "Time and a few kind words." I still love that phrase to this day.

The craziest thing that ever happened to me while squirrel hunting involved a squirrel running up a large tree before I could get off a shot at him. There was a large limb leaning against the tree, and I could picture him running down that limb to get away later, when he decided to come down. Well, after about a five-minute wait, he came down all right, like a bullet. He hit the limb at full speed, as I fired off three quick rounds from my Winchester 12-guage pump shotgun. I missed. He ran about twenty yards away and stopped to rear up on his back legs and made a chattering noise, as if laughing at me. I laughed, too.

As my son, Mark, became a young man, I knew I wanted to share those same moments with him out in the woods. Mark took to hunting like a duck to water. He quickly achieved his youth hunter status, and he and I would spend more time outdoors.

Once, we were squirrel hunting along the hills surrounding Franklin, Tennessee, where the present-day Cool Springs Mall stands. Mark was walking in front of me, carrying a .410 single-shot shotgun, when we encountered a copperhead snake. I told him to back off and give me the gun. I then proceeded to shoot the snake. We imagined that the hills might be covered with snakes and went to hunt elsewhere.

We decided to go deer hunting on Mr. Coffee's farm, just outside of Readyville. He was gracious enough to let us hunt there anytime we wanted. While scouting around, we noticed a couple of his cows had died but didn't think much of it at the time. I let Mark carry the gun, and we approached a bluff area. He was using my Browning 30.06 lever action. After finding a suitable tree for him to climb up in, he did so, and I handed the rifle up to him. Upon hunting about an hour, something let out a scream, like a very large cat. I whispered to Mark that if he saw the cat, shoot him before he got to me. After about an hour of additional hunting, not seeing any deer, or hearing anymore out of the cat, we decided to descend the bluff. We searched for more deer, but to no avail. We called it a day.

The next year, upon going back to Mr. Coffee's, we set up at a tree just below the bluff. Sitting below Mark, leaning up against his tree, maybe a half hour into our hunt, I could not believe my eyes: a large black panther at the top of the bluff. I whispered to Mark to look, but he couldn't get a good view of the cat. After deciding to hike up the bluff, to look for the cat, we reached the summit.

We found paw prints as large as my fist. Once again, we decided to leave, hunting for deer along the way. We spotted one deer; however, the thick brush it was in prevented Mark from getting a decent shot at it. Upon returning to Mr. Coffee's farmhouse, we asked him. "Mr. Coffee, have you ever seen any big cats on your farm?"

"Just our house cats, Bonnie and Clyde," he said.

"This was no house cat. It was a panther," I said.

Mr. Coffee just stared at me, shocked.

Even though Mark and I never went back to Mr. Coffee's to hunt, Mark would later knock down two deer with one single shot. I still don't know how he pulled that off. He has been a skilled sharpshooter ever since. In fact, Mark went on to join the Marine Corp, achieving the rank of sergeant. Mark went on to acquire a forty-acre tree farm in southern Kentucky, which is perfect for deer hunting. Our friend, Gene Dwyer, once bagged a nine-point deer on that property, which made for some good eating.

On one cold day in November, Mark and I were hunting on the tree farm. We decided to do it up right, so we camped out, bringing a grate for the campfire, complete with an old percolator coffee maker. After hunting the first day, with no activity, we decided to settle in around the campfire for the evening. Fixing a breakfast meal instead of supper, we had sausage and eggs and biscuits, with sorghum molasses, and plenty of hot coffee obviously.

Around ten o'clock that night, it began. It started with one howl, from what sounded like a hundred yards out, and then another, within about fifty yards. I knew from the sound it must be wolves. This kept up over the next ten minutes, with the howls getting closer and closer, and beginning to encircle us. At this point, I grabbed my rifle and told Mark that I was headed for the pickup truck.

"Really, Daddy?" he asked.

"Yes, I'm serious." I responded.

He then grabbed his rifle and followed suit. Once safely inside the truck, I called my friend, Dylan, in Nashville, on my cell phone. Asking him to pull up on his computer the sound of a wolf howling, he did and played it for us on my speaker phone. It was the exact same sound we had just been hearing.

After spending about an hour in the truck, and letting the howling subside, we decided to go back to the campsite. We stayed up, drinking a little more coffee, before trying to get some sleep inside the tent in the bitter cold November weather. Having our rifles next to our sides, Mark eventually fell asleep. I could not. I got up in a couple of hours, only to find three wolves encircling the campsite.

Instinctively grabbing my rifle, I whirled around as fast as I could, picking off all three wolves. Mark quickly woke up, amazed at all the commotion. We got up and examined the dead wolves, making certain there were no others around.

We made more coffee and stayed up the rest of the night at that point. The only sound we heard the rest of the night was from the screech owls, as they would get closer and closer to us, evidently intrigued by the campfire. Truly a night to remember.

Another reason to rejoice in the Lord.

Family Matters

As news of my divorce spread, my family responded in different ways. My sister, Susan—being a full ten years younger than me—went on to graduate from a small Church of Christ college with a degree in psychology. After my first mental breakdown, she and her fiancé, Chester, called me at work one day and flat out asked me not to attend their wedding. They gave no reason.

I could only imagine that my recent divorce from Misty, combined with their Church of Christ background, led them to their conclusion. You see, at that time in the Church of Christ, you could not "put away your wife" for any cause other than adultery. The truth is I never cheated on Misty. We just grew apart. They simply assumed too much.

Anyway, I felt like with her degree in psychology, combined with my recent mental breakdown, I would have received a little more empathy from them. Almost thirty years later, Susan asked my forgiveness, which I freely gave. I've since become close to her, Chester, and their children.

There was one other experience with a sibling that took me by surprise. My middle brother, Ray, a successful preacher and Bible teacher at a Church of Christ college, joined us for dinner at our mother's home one evening. Once he arrived, he approached me and called me a loser. I was so taken aback, I couldn't even speak. This coming from a preacher of the gospel, not to mention my own brother. Ray had not even kept up with my life. He had no idea of the four major projects I had been working on and what was involved. We would later make up, and I do forgive him, but there is nothing like a wound from a family member.

Chapter 5
Kentucky Basketball

One could say I inherited my love for the University of Kentucky basketball program from my grandpa Riley. After all, watching him going around with a transistor radio practically glued to his ear, listening to the famed Cawood Ledford announce every game, would have a lasting effect on a young man.

If you're not from Kentucky, you probably never will quite understand what a true college basketball fan is. To say we "bleed blue" is indeed an understatement. Grandpa, of course, knew all the legendary players, like Dan Issel, Cotton Nash, Pat Riley, Louie Dampier, and the rest.

And because Grandpa lived with us from the time I was six until I was fifteen, his love for Kentucky basketball had a big influence on me.

Oh, sure, there's that other university seventy miles across the state that some Kentucky basketball fans tend to hate, but I am not one of them. I admire the record of national championships the University of Louisville has amassed for themselves; I appreciate their record as fellow Kentuckians. Some would say that I'm not a true-blue UK fan, but having witnessed the University of Kentucky's victory over Louisville in the 2012 Final Four was gratifying enough for me. There is something to be said for being there live for the game. I have been blessed enough to have attended five of UK's Final Four appearances. With an overall record of two-and-three, I sometimes wonder if the university would like to compensate me for staying away next time.

Even though the Wildcats went on that year to defeat the University of Kansas Jayhawks in the final game, I am still bothered by what the opposing coach said after the game. As the reporter interviewed coach Bill Self right after the game, he said, "We didn't get beat. We just ran out of time."

Well, I thought, *couldn't you say that about every game? After all, isn't that why they play with a clock?* By the way, that was the year Kentucky had Anthony Davis "the 'Eyebrow" as the superlative shot blocker.

So, when the opportunity came to attend the final four in New Orleans, I jumped at it. Kim and I enjoyed the great Cajun cuisine, taking a bicycle buggy ride from the French Quarters over to a casino and dropping a couple of hundred bucks on blackjack. We also drove over to Biloxi to win a little more money at blackjack.

My next most memorable Final Four included the University of Kentucky's championship game win over Syracuse in 1996 in the Meadowlands near New York City. Kentucky coach Rick Pitino called that year's team the best college basketball team of all time. Being so talented in the starting lineup, and so deep off the bench, few could argue, especially when you consider that Ron Mercer, a freshman from Nashville, would come into the game later, to score twenty points.

I'm convinced there were angels with us on that trip to New York City. First, my second wife and I were walking back to the hotel room in Manhattan, late, and I noticed we were being followed by a group of tall, young men. I say followed because we were not walking fast ourselves, and they made every turn at every corner that we did.

At one intersection, I spotted a police officer across the street. I was hoping he would notice our situation, but he simply turned away as if to say, "You are on your own."

When we reached the hotel and entered the lobby, the group of twelve or so followed suit. We walked to the elevator, and got on, at which point one of the group walked in front of the elevator and, rather than entering, just stared at us. I stared right back at him, until the elevator doors closed. We were spared. That could have easily ended up being a mugging or worse.

The second close encounter occurred immediately after the championship game. We took the bus ride back from the Meadowlands to the bus station in New York City. Upon arriving at the station, we were at the tail end of a group of about six or seven fans. Suddenly, from around a corner, I noticed a small group of people dressed in dark clothing, coming out of the shadows toward us. Instinct told me to reach for my back pocket and retrieve my cell phone (one of the larger ones of that day). For some reason, the group turned and ran. Fear of a cell phone? I doubt it. Perhaps they saw something supernatural, just like the gang at the hotel. I choose to believe we were being divinely protected on that trip.

My final reference to the Syracuse game does not involve the supernatural, just simply an event involving fans before the game.

We arrived early for the game, very early. There were barely any fellow blue-clad Kentucky fans. We had lucked into some seats that put us about twenty rows up from the court. We had bought our tickets from some UK students who had to get back to school for final exams. That left us with only tickets to the championship game.

We made friends with a couple of suit-clad gentlemen sitting behind us, who were from New York City. Then in walks about twenty or so orange clad, and orange-painted Syracuse college fans, running down to their seats, about thirty feet from us. The fans began yelling. "UK. You suck. UK. You suck," and on and on.

I turned to the two men behind us and asked what they thought I should do about this. To which they replied, "Well, up here, we like to say 'Yo Mama' a lot."

I walked over to the group of fans, and as they chanted "UK" I would point to one of them and yell, "Yo Mama." And each time they would do it, I would point to a different fan with my chant. Meanwhile, Kim was pulling at my jacket begging me to sit down. I reluctantly sat down, and at the same time the chant stopped.

Happy Endings

Kim and I had made a pact, having been separated a few months prior to the Final Four, that upon UK achieving the Final Four championship, we would get back together. And upon the sounding of the final buzzer, she jumped in my arms and yelled, "We won!"

In 2011, I had a very eventful trip to the Final Four tournament in Houston, Texas. I was fortunate to be accompanied by one of my father's best friends, Bill Hardcastle, a huge University of Kentucky basketball fan and a retired sergeant major in the United States Army. About halfway down on our drive to Houston, we were cruising along in the fast lane somewhere in Mississippi, when suddenly Bill yelled, "Watch out!" In the road in front of me was an eighteen-wheeler at a complete standstill. At the last second, I was able to swerve into the slow lane and avoid the truck. We both felt relieved.

Then it couldn't have been more than an hour later, when Bill received a call on his cell phone, that his beloved shih tzu dog had suddenly died. She had been his and his deceased wife's, Carol, favorite pet for the past seven years. I could tell this was a tremendous loss to Bill, relating it to the several times I had lost my pets in the past. I asked him if he still wanted to continue to the game, and like a "true blue" fan, he said, "Yes, of course."

Parking at the arena was atrocious the day of the game, and I realized that the long trek was difficult for the seventy-four-year-old Bill. And upon reaching our designated seats up in the rafters, he commented this was not a very good view. I bit my tongue.

Thanks to about a dozen missed free throws by UK, the University of Connecticut Huskies managed to pull off the semifinal upset by one point and advance to the final game.

Still somehow, Bill and I had a delightful conversation on the way back home and knew that UK would be back to fight another day. I will always have a special place in my heart for Kentucky basketball.

Chapter 6
Kim, Dogs, and UFOs

I first met Kim when I was working as an accountant in Bellevue just outside of Nashville. She came to our offices inquiring about our services, and we hit it off right away.

Kim is a beautiful Asian American woman who was born in South Korea. Her parents moved to Nashville when she was in high school. Her father was a talented physician, and Kim wanted to follow in his footsteps, so she entered the field of dentistry. Her practice was close to my accounting office. She is a soft-spoken, humble person who is easy to be around.

As Kim and I got to know each other, we discovered a common interest in gambling from time to time. We have had our share of good luck with the horses and with the casinos—in particular, blackjack and Mississippi stud poker in Tunica, Mississippi, just south of Memphis. When Dueling Grounds racetrack opened in southern Kentucky, we hit the Pik-6 twice, with total winnings of $8,000. To win the Pik-6, you must pick the winner of all six races. Talk about extremely long odds.

Personally, my favorite table game is blackjack. I've had runs where I converted one thousand dollars in chips into four thousand dollars in chips. I've experienced the same success in Mississippi stud poker, but to me, it's just not as fun as blackjack.

Now, Kim—who has similar luck in stud poker—pulled off the "big hand" one day, drawing a royal flush and winning sixty thousand dollars. Quite a haul!

For a brief, two-year period, I worked nights at United Air Lines in Nashville. One of the benefits included free air travel, so

Kim and I decided to fly first class to New Zealand for a two-week vacation. We had a blast. Able to recline—even sleep—in first class for the long flight was delightful, not to mention the hot meals and champagne. The people in New Zealand, on both islands, were very hospitable. Driving on the other side of the road on the southern island, however, proved to be a challenge. One day, while rounding a bend in the road, a tractor trailer appeared seemingly out of nowhere, barreling down on us. I spotted a safe place on the side of the road just in case I had to ditch the car, but at the last millisecond the big truck swerved, and we were spared. Rejoice in the Lord, indeed.

On one of our more recent trips, Kim and I flew to Phoenix, Arizona, and rented a vehicle and made our way all the way up to Yosemite National Park in California. We were looking forward to taking in the massive redwood trees. But upon reaching the parking area, we noticed a sawhorse sign that said, "Trailhead closed." I stated to Kim that I had come all the way from Tennessee, and I was going to see me some redwoods. There were no other cars in the parking lot. So we set down the quarter-mile trek.

Once we found the redwoods, we marveled at the sheer size of them and took some pictures. We were about to leave, when we noticed how deathly quiet it had become in the forest. Immediately thinking "bear," I knew to make some noise, so as not to startle him, so I started singing Lee Greenwood's "God Bless the USA" Then within about one hundred yards of the trailhead, we noticed a man who appeared to be Asian walk across the trail. Upon reaching the parking lot, there still were no other cars around. It was getting close to sunset, so we decided to head toward our hotel near Lake Tahoe. We had a great trip.

When I got back home, I scanned through all the newspapers we saved while we were gone. I noticed an article about a sniper on the loose in Yosemite. Evidently, an Asian concessionaire company had failed to get the concessions bid at Yosemite National Park, and one of the employees went a little crazy, trying to pick off the tourists. Why he didn't get us, we may never know. Just the fact that he could have noticed that Kim was Asian could have been enough to spare us. Rejoice in the Lord!

Dogs

Something else Kim and I had in common was a love for animals. Kim's soft spot for dogs caused her to rescue a three-year-old dog named Coal. Coal is a half-pit bull, half-Labrador retriever, solid black except for a white patch on his chest, and stands about eighteen inches tall. I instantly loved this dog, and the dog loved me.

It seems Coal lived off the land in his early years, scrounging up the occasional squirrel or rabbit. That's my guess, anyway, because every time we see a wild animal, Coal perks up and goes into action. On one such occasion, we were riding through the valley headed toward our parking spot, with the windows rolled down. Upon seeing a squirrel, Coal jumped out the window and chased the squirrel. However, one of his back paws got struck by the car tire and bloodied him a little bit. Upon arriving home, I dressed his wound. He would later be just fine, but this broke him of chasing squirrels from that point on.

I have seen him chase turkey, deer, and even snakes. Coal attacked snakes on two separate occasions, killing both. One was an adult copperhead, and although finding it a bit humorous as two halves of the snake floated down the stream, I knew that Coal was in grave danger of being bit. So, I got permission from Mr. Richardson, the owner of a farm near my new home, to run Coal at his place.

On one cloudy day, with storms threatening, Coal was enjoying a hunt. He was on the other side of a large, weedy brush pile from me, when I noticed a tuft of weeds fly up in the air from his direction. Going to see what all the fuss was over, I got there just in time to see him land a groundhog. He was so proud and wanted to carry it to the car to take with us. Coal is not a huge dog, but the groundhog was every bit as large, and heavy, as Coal.

Having taken him almost religiously to "the valley," which is a beautiful site on Big East Fork Road, near Leiper's Fork, Tennessee, for his twice-weekly runs, he loved even the mention of the valley. Our excursions into the valley have become special to me. I now make it a point to pray fervently and get closer to God as we move through the scenery.

UFOs

After ten years of marriage with Kim, we had reached a point where we needed to separate for a short period of time. During that separation, I must admit I would occasionally visit a few bars and night clubs in downtown Nashville, which I would later regret and repent.

But one night, leaving well after 1:00 a.m., I drove toward Trinity Lane going north on I-65 and headed to my rented garage apartment in Goodlettsville. The night was chilly, and there was hardly any other traffic on the road. As I got closer to Trinity, I noticed an extremely large, dark object in the sky shaped like a triangle. It was just hovering there, at about three hundred feet above the ground, and looked about the size of two football fields. Keep in mind, even though I was out late, I had not been drinking. I rolled my windows down to listen to it, but there was no sound coming from it. Then, it started to veer toward the east slowly, and I decided to follow it, so I took the next exit.

And then suddenly, as if it had noticed me, it sped off toward the east, in what had to be approaching the speed of light. It was out of sight within milliseconds.

Now, what all does this have to do with the spiritual realm and our belief in the Bible? I believe that our God is omnipotent, all powerful. He can indeed create virtually anything he thinks of, but that doesn't mean he has yet informed us of all his majestic works. I believe in heaven we will learn of many more mysteries.

This is the only time in my life when I have seen a UFO. That experience was enough to cause me to be a believer in UFOs. I now pay attention when others try to tell me their UFO sightings. And I would soon have that chance.

Kim and I decided to take a short trip to Fall Creek Falls State Park in Van Buren County, Tennessee. The park is home to the tallest waterfall east of the Mississippi River, although not as majestic as Niagara Falls. The park also had a lodge, with restaurant, cabins, and a golf course. We stayed at a friend's cabin, which was very close to the park. We ate our meals at the lodge, viewed the falls, and played a couple of rounds of golf.

One couldn't help but notice all the UFO magazines lying around the cabin. I picked up a couple and thumbed through them. I thought of the experience I had with my encounter on Trinity Lane. One evening, after leaving the lodge, having enjoyed their buffet meal, as we approached the exit to the park, we encountered a man with a broken down white pickup. There was something about the gentleman that I felt I could trust, so we stopped to offer our assistance. He introduced himself as Jack Cummings and upon asking how we could help, he stated that he could use about a two-mile ride and from there, he could get the help he needed for his truck. We said sure, and he jumped in the car, very grateful.

As we drove the two miles, I happened to ask him if he had ever noticed any UFOs in the area. His response shocked us.

"A couple of years ago, when I was in my last year as Sheriff of Van Buren County, virtually every night we were receiving calls about UFO sightings across the county. Everyone that we could possibly have time to respond to were legitimate unidentified flying objects."

"Have you ever seen any extremely large UFOs?" I asked.

"Yes, usually in a triangular shape," he said.

White Bridge Road

Norm Swift, a businessman in Old Hickory, Tennessee, hired me to raise six million dollars for a tech venture. I expended many hours on that project, helping with the business plan, the financial projections, and presenting the deal to prospective investors.

Norm was a huge cigar aficionado. Many times, we would sit on his back deck, with his wife Jenny, just relaxing and enjoying a cigar, a beer, or both. On one such starlit evening, he led me out into his yard and pointed to a star and then said that star served as the sun for his home planet Kwiznar. Yes, his "home planet." He then shared with me the details of how he was sent to Earth on a fact-finding mission, and if I ever discussed this with anyone, something strange would happen to me. Not death, but more like a sign to remind me not to talk about this topic. He then gave me an example of a couple of friends that he had confided in. Evidently, they were sitting in their living room one evening and suddenly heard the sound of a vehicle crashing through their living room wall. Nobody was hurt, not even the wall, just a jolting reminder not to discuss it with anyone else again.

A few months later, I was watching the college football national title game, at the home of my good friends the Schmidt brothers, Robert and Doug. At halftime, I relayed Norm's story to them. Although they found it interesting, they were skeptical. After the game, I took my usual journey home, going down Knob Road to White Bridge Road. Upon turning onto White Bridge, it was as if something had hit my car and exploded in a ball of fire across my front bumper and hood. I stopped my car, got out, and called 9-1-1 on my cell phone. Walking back to the point of impact, I noticed three men dressed in black T-shirts and long black pants. But as I got closer to them, they simply vanished. In the street, there were two five-gallon buckets, one of which was still intact, and read "explosives." Had I received the sign Norm spoke of? And who were the three men? Angels sent to save me from an impact which could have been much worse? The remaining bucket that I had not hit was upright, as if placed there intentionally by someone who knew my normal course home. An enemy, per chance? At any rate, I thank God I was unhurt. Rejoice in the Lord!

Chapter 7
Ghosts

The bloodiest battle of the Civil War was fought in Franklin, Tennessee.

With over 10,000 casualties in one day, the Battle of Franklin was fought all throughout the city of Franklin and several miles beyond. It was an all-encompassing battle. Evidently, the South's General John Bell Hood decided to ignore the failure of Pickett's Charge at the Battle of Gettysburg and ordered an all-out frontal assault upon the Union army. It, like Pickett's Charge, was a total disaster.

Because I grew up in Kentucky and Tennessee, I was very familiar with many of these historical sites. In fact, I am a direct descendant of soldiers from the North and the South. I have two great uncles—a captain and a lieutenant—who fought for the South under the great General Nathan Bedford Forrest. They achieved specific notoriety for capturing six hundred Northern troops at the Battle of Vicksburg and at the Battle of Brice's Cross Roads in Mississippi. At Brice's Cross Roads, Forrest, with a force of only four thousand, completely routed his opponent of seven thousand.

I also had a relative that fought for the North, Major William McKinley. He fought under General Rutherford B. Hayes, both of whom went on to become president of the United States. McKinley was not a graduate of West Point, rather, he rose from the ranks as private to major; two of his promotions were attributed to valor under fire.

Later in life my son, Mark, and I took a vacation to Virginia with my friend, Robert, where we toured twelve Civil War battlefields. We also made an excursion into Pennsylvania to take in Gettysburg, one of the most spectacular battlefields in the United States.

That night we stayed at the KOA at Harpers Ferry; it was a foggy night. While driving to our designated camp, we saw a ghostly figure dressed in uniform, complete with rifle, walking along slowly, almost as if on sentry duty. We passed it off as mere coincidence and turned in for the night.

In the middle of the night, I had the urge to pee, and rather than walk all the way to the restrooms at the other end of the campground, I decided to just "go around back." Upon heading around back, I soon noticed a shadow of what appeared to be a young man. The shadow was about forty feet from me and walking in the same direction. When I looked around to see who was casting the shadow, based upon the location of the light source, there was no one. I looked back, and it was gone.

On this same trip, we toured the Battlefield of Petersburg in Virginia. During this battle, General Grant had General Lee pinned in with what is called a siege, where the attacking army has the defending army surrounded. The attacking army attempts to starve out the defending army, hopefully forcing them to surrender.

The three of us had spent the better part of the day touring Petersburg, seeing the infamous Crater, a part of the battle that had gone badly wrong when dynamite in a tunnel was sorely mistimed. There is an unwritten rule within the National Park Service, as concerns the battlefields in particular: All visitors must leave the park grounds before nightfall. During the filming of the movie, *Gettysburg,* there were some reenactors who were straggling off the battlefield, and they began to get pelted with "ghostly" mini-balls. Mind you, not enough to injure them but certainly enough to get their attention.

Just before twilight, we noticed the Petersburg park service personnel leaving the visitors center. We knew we should leave soon. Robert had already gone to the car, about thirty yards from me. Mark was on the other side of some hedges, about twenty yards from me. I decided to examine one of the old mortar cannons that they usually mounted on train cars.

About that time, there it was—a bugle sounded, and someone called out, "Charge!" That was followed by three blasts from a cannon, at very close range.

Boom! Boom! Boom! Mark came running around the hedges toward me, yelling. "Daddy, did you hear that?"

"Yeah, I did. Let's get out of here!" I said.

Upon reaching the car, we asked Robert if he had heard it.

"Hear what?" he said.

After touring about a dozen Civil War battlefields in Virginia, Robert, Mark, and I were finally traveling home. Robert was driving, and it was nighttime, as we approached the town of Boone, North Carolina. Now Robert was known for having sleeping problems, and this drive would show us just what that meant. As we continued driving through the night on those curvy Blue Ridge Mountain roads, I looked over at Robert and noticed his eyes were closed. I did not dare to wake him, for fear that doing so would startle him and we would wreck. Well, he drove the rest of the twenty miles to Boone without incident.

Angels protecting us? I think so.

My favorite ghost story happened while I was living in a condominium complex in Franklin near the historic battleground area.

Whether working as a CPA or when I was a business broker, I tend to get up in the middle of the night, work for two or three hours on the computer, and go back to sleep. This night was no different. I awoke at about two and worked on the computer for a couple of hours. Then I planned to take a short break by playing my harmonica for a little while. I broke into "Dixie," and almost immediately there came a loud shrill scream from down the hall in my master bedroom.

"No!"

I slammed the harmonica down, as the hair rose on the back of my neck. I finally mustered up the courage to venture down the hallway to the master bedroom. Seeing nothing there, I retreated to the recliner in the living room and eventually dozed off.

The next day I explained to a friend what had happened. She was very attuned spiritually and began to convey what she believed took place. She believed it was a very young Union soldier, clad in blue, possibly a drummer boy. He was killed in the Battle of Franklin. She also stated that either he did not appreciate hearing me play "Dixie" or that he thought I was butchering the song terribly.

Finally, there's the story of Ben Wilson, of Goodlettsville, from whom I rented a garage apartment for a while. Ben was a jovial old fellow, a retiree of General Electric.

One morning, my landline phone awakened me, but before I could get to it, it stopped. Then, my pager went off on the table beside me. It was as if someone had smacked the pager, and it flew across the room, landing some twenty feet away on the carpet. I immediately went downstairs and knocked on Ben's door. I asked him if he'd ever had any unusual events occur here and told him what I had just experienced.

"Yes, ever since my mom passed away last year—in the garage apartment—many such events had occurred." He described the sound of someone walking through his house, lights being turned on and off, and even someone knocking at the front door, with nobody being there.

While living there, I typically started my day by reading portions of the famous Christian devotional *My Utmost for His Highest* by Oswald Chambers. One morning, just as I completed reading about *a mighty rushing wind* on Pentecost in Jerusalem, Ben came knocking on my door.

"Did you hear that?" He said. "It was a loud, rushing wind going through the house."

"No, I didn't hear it," I said, "But I was just reading about it!"

Chapter 8
Against All Odds

It was a splendid fall day in October. I decided to go to Barnes & Noble in Cool Springs to peruse their latest list of best sellers.

Upon turning a corner, there she stood, the most beautiful woman I had ever laid eyes on. I mustered up the courage, planning what to say, and approached her.

"Excuse me, has anybody helped you yet?" I asked.

"Oh, yes, thank you," she answered.

Relieved she didn't need any help, I then stated, "Hi, I'm Steve, and I really don't work here. I just wanted to meet you."

She giggled and said, "Hi, I'm Melody, and it's nice to meet you."

We chitchatted a while, long enough to determine ours would be a long-distance relationship with her living in Benton, Kentucky, and me being in Franklin, Tennessee, eventually deciding that we would have our first date on a Tuesday night at the Red Lobster in Clarksville, about halfway between us, and then we said our good-byes and parted ways.

I arrived early as is customary for me—whether business or pleasure—since I wasn't sure exactly how to gauge the Nashville rush hour traffic. I enjoyed a glass of Cabernet while waiting, and when she walked in; she literally lit up the room with her beauty and vibrance.

We greeted each other with a hug and took our reserved table. After ordering our meals, Melody asked if we could say the blessing. We held hands, and both of us began to pray. I conceded immediately and let her finish her prayer. We got a little chuckle out of that. It was that night that I first realized what a strong woman of prayer she was, praying with her whole heart.

Over dinner, we discussed our professional and personal lives. She was forty-six, I was sixty. She was divorced with three children, and I was divorced twice with two children. She was originally from Belize while I was from Kentucky. Melody was in training to be a registered nurse and loved her work. I was a former CPA and currently a business broker. The conversation continued effortlessly throughout the course of the meal and into the night. We had such a good time we decided to meet again. I walked her to her car and gave her a peck on the cheek. We said good night, and I asked her to call me when she reached home safely.

Over the next few weeks, we found ourselves falling in love, seeing each other at least once a week and texting daily.

Steve, I love you. I have been hurt in the past, but I still believe in love, in your love.

I responded. *In a way I wish you could experience for five minutes how much I miss you when we're apart, but then again, I wouldn't want to wish that on anyone.*

You are very important in my life. I would like to share my life with you. I can't wait to see you on Thursday.

Sometimes it will be like us against the world. Hopefully in times like that, our love will grow stronger for each other. I wrote.

My Valley

Eventually, I took Melody out to my favorite spot near Leiper's Fork, the valley. The same place I used to take my dog Coal to run. Melody and I spent many times there, enjoying nature, meditating, and praying. It was there on a beautiful sunny November day that I proposed marriage to Melody. I had gone way into debt buying her the ring that I thought was best suited for her, but that didn't seem to matter. I felt like the happiest man in the world when she accepted my proposal.

However, it was a different story at Thanksgiving, when she and I announced our engagement to my family. Virtually everyone besides my mom and my second oldest brother turned a cold shoulder to us. To say that we weren't expecting this puts it mildly. I got unwanted advice later from even my children, that "You haven't known her long enough," and "I think it looks like you may have spent too much on a ring." Melody didn't handle this well, and I couldn't blame her. After all, it seems like they should be happy for us, at such a happy moment in our lives.

We had decided to refrain from sex until we were married—for spiritual reasons.

However, one night, we experienced some heavy petting. Melody texted me later, saying, *Please forgive me. I know last night I was the one who wanted to be touched by you. I know you didn't feel comfortable doing it. I am sorry. I feel guilty. I won't put myself in that position anymore. I am sorry, Steve.*

I explained to her. *That's okay, Melody, I was a willing participant, too.*

Meanwhile, I had planned to sell my condominium and move to Colorado. There were two problems with that. One, Melody wanted no part of moving to Colorado, or even leaving Kentucky for that matter, due to her having only joint custody of her youngest child. And two, for the homes I had recently viewed in Colorado, I would need to receive a large bonus from one of my business ventures, to afford the down payment. And this bonus had recently been put off for another year. Well, I did go ahead and sell my condominium and attempted to acquire a home in Kentucky, to be a little closer to Melody.

Foggy Night, Valentine's Day

On Valentine's, Melody had to work, so we planned to meet on Monday night in Clarksville. At dinner, we discussed the terrible fog out tonight. After eating, we went to my car and discussed whether Melody should travel back across the lakes in that fog or try to get permission from my mom to stay there with her: Melody in the guest room and I on the couch. We even investigated a motel room for the night. After all, this was by far the worst fog either one of us had ever witnessed.

We decided that I should call my mom. What a mistake. Number one, she had no idea how bad the fog was; and number two, she thought I meant I and Melody would be sleeping together, before I could explain the expected sleeping arrangement.

She simply started yelling. "'I'm mad. I'm mad!"

To which I yelled back. "I'm mad too."

And we both hung up. Melody heard all of this and was very upset, as was I, at my mom. We then checked into the local Econo Lodge for the night. The night was not peaceful. We argued about my mom and her lack of acceptance of us as a couple. Deep down, I really agreed.

At about four o'clock in the morning, Melody had enough and decided to make the trek through the fog and back home. This was the low spot in our relationship. My mom and I made up almost immediately, realizing that her old-fashioned ways had kind of outweighed the gravity of the matter with the dangerous fog that night.

A few days later, Melody had told me that she'll be working all night and please to text her as much as I want to. After texting her several times with no response, I finally sent her a message stating, *What's your problem?*

You're my problem.

Well, did I ever find out over the next few days, just how much I was the problem. Melody had just found out she was entitled to one-half of her ex-husband's military pension, unless she remarried. And since she had worked hard for her share of this pension, she was unable and unwilling to get married. I informed her that I will need to get the ring back from her, to which she replied that she had already spoken to her attorney about that and that she could keep the ring. I immediately got my attorney involved, and he sent a great letter that induced her to return the ring.

Parting Shots

Melody called me on April 8. She stated that she was going through her drawers and ran across the tickets to the Brian Adams concert that I'd given her for Christmas and wanted to know if I wanted to go with her that next Friday. I said, "Sure," and we had dinner at Mere Bulles and enjoyed a great concert together.

She will find happiness again someday, I'm certain.

She's a sweet and beautiful Christian lady and deserves the best.

Chapter 9
Parting Ponderings

Thank you for joining me on this journey. I have shared many stories with you, and I want to end with a few thoughts on the Church of Christ and leave you with one of my favorite spiritual poems that can also serve as a prayer.

After meeting so many good Christians of different denominations over the years, I have come to realize God's house is big enough for all people who love him. Nevertheless, I believe the doctrine of the Church of Christ is probably closer to the truth than any other denomination, and I say that with confidence. I appreciate their stand on baptism and participating in the Lord's supper every Sunday. Today, some Church of Christ congregations are finally easing up on their "no musical instruments in the worship service" policy. I think this is a good development. Are we not judging billions of souls by demanding that they worship without instrumental music? If so, then why don't we take the Bible to a few other extremes, like only reading the Scriptures in the original Hebrew or Greek? Of course we don't, and the same is true for the way we worship. As the Bible says, "…for man looks at the outward appearance, but the Lord looks at the heart." (1 Samuel 16:7, NKJV; biblegateway.com)

Thankfully, however, I am not the judge, only God has that responsibility. I love all my brothers and sisters in Christ and, yes, even my enemies.

Did you know that of all the people who profess to be Christians, only about 17 percent attend church on a regular basis? I know, because this is something I have struggled to do consistently in the past.

I believe we are now closer to the end times than we have ever been before. The Bible teaches us that during the end times, we are to gather together even more often than usual. This is a good thing. I really enjoy the basic tenants of worship service, including edification, the Lord's supper (or communion), the giving of tithes and offerings, and music.

Unlike Oprah, I believe there is only one true path to God. But many people follow what she and many others like her are teaching. Don't let your mind get watered down. Stand firm.

Here are several statements about my faith I have put together over the years. I often quote these aloud. They help me to stay grounded in my beliefs. It has helped me for years, and I believe it can help you too.

I recognize that there is only one true and living God who exists as the Father, Son, and Holy Spirit. He is worthy of all honor, praise, and glory as the one who made all things and holds all things together (the book of Colossians).

- ✓ I believe that God demonstrated his own love for me in that while I was still a sinner, Christ died for me. I believe that he has delivered me from the domain of darkness and transferred me to his kingdom, and in him I have redemption, the forgiveness of sins (Romans).
- ✓ I recognize Jesus Christ as the Messiah, the word who became flesh and dwelt among us (John).
- ✓ I choose to obey the two greatest commandments: to love the Lord my God with all my heart, soul, mind, and strength and to love my neighbor as myself (Matthew).
- ✓ I obey the command to submit to God and resist the devil, and I command Satan in the name of Jesus Christ to leave my presence (Matthew).
- ✓ I obey the command to repent and be baptized, in the name of Jesus Christ, for the forgiveness of my sins, and that I will receive the gift of the Holy Spirit (Acts; Mark).

Finally, I want to leave you with a poem I wrote. I hope it inspires you.

How Are *You*, Lord?

Here I bow before Your grace
and see you face-to-face
as You hold me in Your loving arms,
keeping me free from all pain and harm.
I take shelter beneath the eagle's wings,
and through all Your awesome glory I sing.

How are You my Lord Jehovah?
for I do care about You, as You care about me.
I know the struggles against evil should make You weary,
though believing that You shall always prevail,
Just as I know by faith that Jesus did hang
on that tree by those nails.

Realizing that You're Omniscient.
Realizing that You're Omnipotent.
And realizing that You're Omnipresent.

I still care about You, and wonder,
How are You, Lord?
Oh, how are You, Lord?

JESUS
Beloved Jesus of Nazareth,
who made that lonely trip to Golgotha's Hill,
the tomb could not hold you for long,
for you remained my Savior still.
It is through Your ultimate sacrifice,
that I can even have a glimpse of Eternity's thrill.

How are You my Lord Jesus?
for I do care about You; as You care about me.
Knowing the struggles against evil should make You weary,
though believing that You shall always prevail.
Just as I know by faith that Jesus hang on that tree by those nails.

Realizing that You're Omniscient.
Realizing that You're Omnipotent.
And realizing that You're Omnipresent.
I still care about You, and wonder
How are You, my Lord Jesus?
Oh, how are You, my Lord Jesus?

HOLY SPIRIT
O, blessed Holy Spirit,
dwelling within my very body as Your temple,
and protecting my very soul.
interceding for me in prayer, and
even praying on my behalf,
Your love for me clearly shows.

How are You my dear Holy Spirit?
for I do care about You; as You care about me,
I know the struggles against evil should make You weary.

Though believing that You shall always prevail,
just as I know by faith that Jesus hung on that tree by those nails.

Realizing that You're Omniscient.
Realizing that You're Omnipotent.
And realizing that You're Omnipresent.

I still care about You, and wonder
How are You, O Holy Spirit?
Oh, how are You, O Holy Spirit?

Chapter 10
That's What God Does

If there was ever a time when I (Bobby) questioned where God was, it was on September 11, 2001. On that day, Americans faced the worst surprise attack since Pearl Harbor. We were devastated. Normally my faith is strong, but like so many others on that day, I was struggling. We were all asking, "Where was God?"

After a while, however, I began to hear stories—miraculous stories—from survivors who were somehow delivered from the devastation by the mighty hand of God. And that's when it hit me. God was right there in the middle of tragedy saving lives. He was not the author of the tragedy, rather he was the savior during the tragedy. That describes God's nature throughout the Bible. That's why I think it is important to share with you some of the stories that restored my hope and faith in God during tragedy.

My faith was restored when I heard how Stanley Praimnath's life was spared when United Airlines Flight 175 crashed into the South Tower of the World Trade Center. I thought, *Yes, that is what God does.*

When I heard about the men from Ladder Company 6 who stopped to help Josephine Harris—and perhaps more importantly refused to leave her after she collapsed on the fourth floor of the North Tower—the firemen instantly endeared themselves to God and to me. I believe, because of their righteous actions, God built a shelter around that little group of humanity between the second and fourth floors of Stairwell B so they would survive when the North Tower finally fell at 10:28 a.m. of September 11, 2001. *That's what God does.*

When Genelle Guzman—a woman who had turned away from God when she was young—found herself buried in the rubble on 9/11, she asked God for a second chance. God granted it. With only one hand sticking out from the rubble, Genelle cried out for help. Suddenly a man named Paul grasped her hand and said he would not leave until rescue workers arrived. Nor did he. When the rescue workers arrived, they pulled Genelle from her hellacious entrapment and celebrated. Ms. Guzman asked about a man named Paul, but the rescuers said there wasn't anyone named Paul nearby. As the last known survivor pulled from the wreckage of the fallen towers, Genelle Guzman had been trapped for twenty-seven hours. I just know in my heart that the man named Paul was her guardian angel, and once his job was finished, he left. *That's what God does.*

God always has an answer for every problem we will face. So, when the 2,700 employees of Morgan Stanley/Dean Witter needed someone to guide them to safety on that terrible day in September of 2001, God gave them Rick Rescorla. Because of his heroic actions, all but six of the 2,700 employees were saved. Rick was last seen going back up into the South Tower to help save others. Rescorla would be one of the six who perished.

Rick's heroic deeds that day remind me of the Scripture, "For God loved the world so much that he gave his one and only Son, so that everyone who believes in him will not perish but have eternal life" (John 3:16, NLT). Now, I'm not saying that Rick Rescorla was the same as Jesus. Nobody could be. Jesus was the Son of God. I'm saying that it's God's love that gives us hope in time of despair. It's God's love that gives us encouragement when we are faced with doom and destruction. It's God's love that rescues us in our time of need. And it was God's love that drives men like Rick Rescorla to put their lives on the line, in order to save others. *That's what God does.*

When I heard these stories, I was convinced God was at work on September 11, 2001. When I saw how God was at work on 9/11, I looked back to the Bible and realized God has made it a point to save and deliver his people from destruction. Whether it was Moses and the children of Israel facing a Red Sea with no way out, Joshua facing the walled city of Jericho, or the prophet Daniel spending an evening in a den of lions, God has a way of delivering his people from peril. I am convinced he does the same today for his children. These examples from 9/11 are among the best our generation will see. I hope they inspire you as much as they have me.

Chapter 11

Where Was God on September 11?

Stanley Praimnath

According to the 9/11 Commission, Stanley Praimnath (pronounced: *pray-aim-neth*) is the only known survivor from either of the impact zones at the World Trade Center towers. Stanley worked as an executive for Fuji Bank on the eighty-first floor of the South Tower of the World Trade Center (WTC2). The second tower was the second building to be hit but the first one to collapse. He was one of only thirteen survivors from, at or above, the impact zone where the plane hit the South Tower of the World Trade Center on 9/11.

Born in Guyana, a small country on the northeastern edge of South America, Praimnath came to America in 1981. As a young man, his mother insisted he attend church, but as a teenager Praimnath stopped going. After he reached the United States, however, Praimnath made a decision that would have a profound effect upon his life. He started going to church again.

Praimnath was hired by Fuji Bank, and by 2001 he had risen to the rank of assistant vice president. He ran all of Fuji Bank's loan operations on the eighty-first floor of the World Trade Center.

September 11, 2001, began like any other day for Praimnath. When Stanley arrived at the World Trade Center, he took the elevator up to his office on the eighty-first floor. While he was in the elevator, at 8:45 a.m., Tower One was hit by the first plane. However, Praimnath didn't see or hear anything. As soon as he got to his desk, he received several phone calls from his family. But none of them told him what had happened. When he hung up, he looked out the window, at Tower One. He could see it was on fire.

"I saw fire falling through from the roof," Stanley said.

Praimnath decided it was time to leave.

Along with a young temporary employee named Delise, Praimnath got on the elevator and went down to the seventy-eighth floor to the Sky Lobby to catch the elevator to the bottom floor. The company's president, CEO, and human resources director, as well as two other men, got on the elevator and headed down to the concourse level of Two World Trade Center.

When they reached the concourse level, however, a security guard stopped them and told them to go back to their office. He said that an accident had happened at the other building and it didn't affect them at WTC2. In all fairness, this was probably the best advice that could have been given at that time. Until the second plane crashed into the South Tower, most people probably thought that it was just a terrible accident. And, with all the falling debris, fire, and, yes, even bodies of those who were jumping off the North Tower, going outside did not seem like a wise choice.

As they headed back to their offices, Praimnath told Delise that she could go home. He could tell that she was upset. The company's president and CEO got off on the seventy-ninth floor while the human resources director went up to the eighty-second floor. Praimnath would never see them again.

The phone was ringing when Stanley got to his office. It was a woman, from Chicago, checking on him. He told her everything was okay. As he was talking to her, however, he looked out his window toward the Statue of Liberty and saw the second plane, United Airlines Flight 175, heading directly toward him.

Immediately, Praimnath said, "Lord, I can't do this—you take over," and dove under his desk.

Praimnath later said, "I knew beyond a shadow of a doubt that the Lord was going to take care of me once I got there."

You see, Stanley's Bible was on top of that desk, and his faith was so great he knew he didn't have to worry. Just like Daniel in the lions' den and the Roman officer who asked Jesus to save his servant, faith is everything. Jesus once told his disciples, "If you had faith even as small as a mustard seed, you could say to this mountain, 'Move from here to there,' and it would move. Nothing would be impossible" (Matthew 17:20, NLT)

The plane slammed into the seventy-fourth thru eighty-fourth floors of the South Tower at 9:03 a.m. It was traveling well over 500 miles per hour when it hit, and it immediately exploded.

According to Stanley, "If you look at the video, the plane is coming straight in toward the building, and at the last minute it makes a tilt." Praimnath believes that God turned his plane as a result of his prayer. So do I!

A huge fireball engulfed the entire floor. The walls were flattened, and furniture was shredded and strewn all over the place.

"The only desk that stood firm was the one that I was hiding under because my Bible was on top of that desk," said Praimnath.

Fire was everywhere, except in Stanley's office. He could see the tip of the airline wing in the doorway to his department. Praimnath was trapped, buried under debris, in imminent danger, and yet he never lost faith in God.

"Lord, you take control, this is your problem now." He recalled praying.

Praimnath pulled himself out of the rubble and began moving toward safety. It was totally dark. Stanley couldn't see anything. He moved by feel alone.

In desperation, Stanley cried out. "Lord, send somebody to help me. I don't want to die. What will happen to my wife and two kids?"

At that same moment, Brian Clark, an executive vice president with Euro Brokers, was making his way down a stairwell with six other people. When they reached the eighty-first floor, a woman told them the remaining portion of the stairwell was blocked and their only choice was to go back up. While the group was talking about what to do, Clark heard Praimnath's screams for help, while the others turned around and went back up.

Clark, who as a fire warden on his floor, had a flashlight and went to look for Stanley Praimnath, while the others went up and perished. Praimnath kept praying for help. He knew he couldn't do it alone. And then, up ahead, he saw a light. It was Brian Clark and his flashlight.

Praimnath said his first thought was, "This is my guardian angel. My Lord sent somebody to save me!"

But, he wasn't out of the woods, yet. After battling his way through a mound of debris, as well as the wreckage of three departments, there was still one wall left between Stanley and his pathway to safety.

Finally, with a little more prayer, and with Clark urging him on, Stanley was able to muster enough strength to batter his way through the wall and into Stairwell A. This was the only stairwell of the South Tower to withstand the impact of the crashing plane.

Clark looked at Praimnath and said, "I always wanted a brother. I found one today." But they weren't out of danger yet. The building was still on fire, and unbeknownst to them, the South Tower was going to fall in less than an hour. WTC2 collapsed in just fifty-six minutes, well before the North Tower fell, and this was despite the fact that United Airlines Flight 175 hit the building seventeen minutes after American Airlines Flight 11 smashed into WTC1.

With eighty-one floors between them and the outside world, the two men started walking down. As they made their way down those stairs, they saw many terrible sights. They saw more when they reached the ground. But the worst was to come. After what must have seemed like an eternity, the two newfound brothers finally reached the concourse level. The firemen—many who were still going up into the building—told them to run as far away from the building as they could. Running through the flames and dodging debris, they slowly made their way through what resembled a war zone. Broken glass and twisted steel rained down, as did the bodies. It's been estimated that more than two hundred people, faced with the terrible choice of burning alive or taking a one thousand-foot plunge, jumped to their deaths. Praimnath and Clark heard the eerie thud as those bodies hit the ground.

The two men finally reached Trinity Church about two blocks away.

"As soon as I held onto the gate of that church, the building collapsed," said Praimnath.

As he watched, in horror, the South Tower, of the World Trade Center, imploded with each succeeding floor crashing down on the one beneath it. Rather than an explosion, the scene unfolded as if one were watching a controlled demolition of a high-rise building. Rather than blowing outward, everything fell in place. A large cloud of smoke quickly engulfed everything and everyone. Praimnath and Clark were separated and would not see each other again, on that fateful day.

Stanley was able to catch a ride with a truck driver and made his way to his wife's office only to find that she had already left. He finally made it home, much later, to embrace his wife, Jennifer, and two daughters, Stephanie and Caitlin.

Stanley Praimnath is convinced that the mighty hand of God turned that plane and that's why he survived. As Stanley thanked God, he told him that whatever he did, "It would always be for his glory."

Chapter 12

Where Was God on September 11? Josephine Harris and Ladder Company 6

The morning of September 11, 2001, was a welcome change in the weather to fire captain John "Jay" Jonas. As the leader of New York Fire Department Ladder Company 6 based in lower Manhattan, Jonas's shift began the night before when he reported for the night shift. The brave men of this company, located on Canal Street in Chinatown, usually fought local fires in nearby tenement buildings. Very seldom did the men of Ladder 6 have to enter the city's skyscrapers.

According to Jonas, the evening of September 10 had been a very stormy night. So, when the clear skies of September 11 arrived, he said they were beautiful. Other members of Ladder Company 6 were driver Mike Meldrum, his tillerman, Matt Komorowski—who drives the rear of the two-section hook-and-ladder truck, steering the rear wheels through the narrow confines of New York City—Tom Falco, Bill Butler, and Sal D'Agostino, the company's roof man.

Josephine Harris worked as a bookkeeper for the Port Authority of New York and New Jersey. On 9/11, she was working on the seventy-third floor of the World Trade Center's North Tower.

With his shift ending, Komorowski was out front getting ready to go home when he saw American Airlines Flight 11 crashed into the ninety-third thru ninety-ninth floors of the World Trade Center's North Tower at 8:46 a.m. The station watchman also saw the impact and called out over the intercom. "A plane just hit the World Trade Center."

According to Jonas, they heard what sounded like a low-flying airplane and then a very loud boom. Within minutes, the members of Ladder 6 parked their fire truck in front of the towers. Debris rained down as they hurried into the lobby of One World Trade Center.

While waiting for their assignment, the second plane, United Flight 11, crashed into the other tower.

Captain Jonas said, "Oh my God, they're trying to kill us."

Gerry Nevins, one of Jonas's friends, said, "We're going to be lucky to survive this."

A fireman's job is to rescue people in a burning building. But Captain Jonas knew they were going to have to climb a long way (eighty to ninety floors) just to reach the fire. And they would have to lug about one hundred pounds of gear to do this. So, he knew they would have to pace themselves. Climbing up Stairwell B, the only stairwell that went all the way to the ground floor, the men stopped every eight to ten floors to rest. When they reached the twenty-seventh floor, however, the men felt the entire building shake. Unbeknownst to them, the South Tower had just collapsed, and they had less than thirty minutes before their own building would fall.

"I knew what that meant," Jonas said. "The sister building to the one that I was in has just collapsed and my building got hit sooner. So, I just started thinking we're not going to make it out alive."

Captain Jonas made the tough decision to abandon the building. He knew if the other tower had collapsed, then it was just a matter of time. Jonas did not find out until later that the decision to evacuate had already been broadcasted. They just didn't hear it.

Somewhere near the twentieth floor, they found a middle-aged woman who was completely exhausted. Her name was Josephine Harris, a fifty-nine-year-old bookkeeper from Brooklyn who had been struck by an automobile. Harris had already descended multiple flights of stairs and was fatigued. She had not fully recovered from her car wreck, and she was feeling it. She had managed to walk down, by herself, from the seventy-third floor. She just couldn't go any further. With Bill Butler, Ladder 6's ironman (and strongest member), helping Ms. Harris, the little troop began climbing down the last of the stairs.

"I could hear the clock ticking in the back of my head," Captain Jonas said. "I'm thinking, 'C'mon, c'mon. We've got to keep moving'."

The small cavalcade of firemen, with their charge, Ms. Harris, walked down, literally one step at a time. Many others passed them and kept on moving until they were out of sight, but none of the hard chargers of Ladder 6 would think of leaving Ms. Harris. Somewhere between the fourth and fifth floors, Josephine Harris collapsed. She could go no further. She pleaded with the firemen to leave her, to save themselves. But they refused to leave her. Tom Falco, the can man of Ladder 6, grabbed her other arm, and with Butler on one side and Falco on the other, they prepared to move on. Captain Jonas hurriedly searched the fourth floor for a chair that they could carry her in. However, with time running out, he gave up his search and returned to Stairwell B. It was at this time, 10:28 a.m., September 11, 2001, that the tower began to collapse.

As each succeeding floor piled on to the next, a huge rush of wind cascaded down the elevator shaft. Captain Jonas said, "There was unbelievable noise." He thought, *I can't believe this is how it ends for me.*

The men were spread out in the stairwell between the second and fourth floors. Matt Komorowski was last in line, and yet in no time he found himself in front of the group—the rushing wind had picked him up and thrown him down two flights of stairs. In addition to Josephine and her six heroic saviors, there were four others trapped in this apparent oasis among the twisted beams and girders and concrete that used to be a stairwell. Port Authority police officer, David Lim; Fire Department battalion chief, Rich Picciotta, of the Tenth Battalion; and two other firefighters found refuge in their little sanctuary of safety. They were battered and bruised, but all of them were alive.

After several hours, the smoke and dust began to clear. Then something eerie occurred. A shaft of sunlight shone down on them. It dawned on them that 106 floors had been above them and now there was just sky. Only after the firefighters emerged did they realize the ghastly enormity of what had happened. Somewhere in the ruins was their flattened fire truck.

Fellow firefighters were finally able to rescue them. Ms. Harris was carefully taken out in a rescue basket. To this day they still can't be sure why or how everything happened, only that it did.

According to Matt Komorowski, "There's no reason we should be alive. We all thought she (Harris) was going too slow," he said, "But she had the perfect pace."

"It was a freak of timing," Captain Jonas said. "We know the people below us didn't fare well. Above, to my knowledge, none got out. God gave us the courage and strength to save her, and unknowingly, we were saving ourselves."

Chapter 13

Where Was God on September 11? Sergeant John McLoughlin and Officer William Jimeno

While Josephine Harris and the men of Ladder Company 6 were going through their three-hour tenure in the twisted remains of Stairwell B, Sergeant John McLoughlin and Officer William Jimeno, a rookie, of the Port Authority of New York and New Jersey were trapped in their own hellacious bubble of life support.

After the first plane struck the North Tower, the forty-eight-year-old McLoughlin, a twenty-one-year veteran of the force, along with rookie William Jimeno joined about twenty other officers and took a bus to the World Trade Center in a rescue effort. One of those officers was Jimeno's friend, Dominick Pezzulo. They were joined by two other Port Authority police officers, Antonio Rodrigues and Chris Amoroso. The five officers were located in the concourse just outside the South Tower when the building gave way.

McLoughlin, the senior officer, quickly led the men to a nearby freight elevator. Two of the officers, Antonio Rodriquez and Chris Amoroso, were killed when the concourse collapsed on them. McLoughlin, Jimeno, and Dominick Pezzulo were spared as the freight elevator (which is stronger than regular elevators) was able to withstand the collapsing steel and concrete. Three officers were managed to survive.

Although McLoughlin and Jimeno were trapped under the debris, Pezzulo was able to get free and had actually started trying to rescue his friend, Jimeno. The collapse of the North Tower twenty-nine minutes later (10:28 a.m.), however, caused the rubble to shift, and Pezzulo, who had survived the collapse of the first building, received mortal injuries.

Jimeno would still be trapped for ten hours after Josephine Harris and the fire fighters from Ladder 6 were rescued. And it would be another nine hours before Sgt. McLoughlin would see daylight.

When Jason Thomas, a twenty-seven-year-old former sergeant, found out about the attacks, he got into his Marine Corps uniform and headed toward the World Trade Center. He had just parked his car when the second tower (the North Tower) collapsed at 10:28 a.m.

Shortly after watching the attacks, from his office in Wilton, Connecticut, forty-three-year-old Dave Karnes, a former staff sergeant, put on his Marine Corps camos and drove all the way to New York City. Arriving at approximately 5:30 p.m., the twenty-year-year infantry veteran joined the twenty-seven-year-old Thomas, in a search looking for survivors.

Climbing over mountains of debris and dodging still glowing pieces of hot metal, the two Marines kept yelling over and over. "Is anyone down there? United States Marines!"

The sun had fallen, and darkness had crept in before they finally heard a faint cry for help. They quickly found an engineer that had a flashlight. After talking to the lost police officers, the engineer went for help and brought back members of the fire department. It took three hours for them to dig out Officer Jimeno and another nine hours to rescue McLoughlin, but they were finally on their way home, although it would be months and many surgeries before the two police officers would be able to go home. In fact, McLoughlin's injuries were so severe that he had to be put into a medically induced coma for six weeks. They did survive to live another day.

Both of our Marine heroes spent several days at the World Trade Center site before leaving. Each one went back into the service. Thomas joined the United States Air Force, while Karnes reenlisted and spent seventeen months, in the Marine Corps Reserve. He did two tours in Iraq.

Chapter 14
Where was God on September 11? Genelle Guzman

While a movie (*World Trade Center,* starring Nicolas Cage and Michael Pena) was made about the rescue of the two Port Authority police officers, they were not the last survivors of the World Trade Center. That distinction goes to thirty-year-old Genelle Guzman, a Port Authority secretary who worked on the sixty-fourth floor of the North Tower.

On September 11, 2001, Guzman, a transplant from Trinidad and Tobago—located just a few miles off the northern coast of South America—was at work on the sixty-fourth floor of WTC1, the North Tower of the World Trade Center. Although she had grown up with a religious upbringing, Genelle had turned away from God many years before that fateful day in September of 2011. According to Guzman, she had done whatever she wanted. She drank when she wanted to. She partied when she wanted to. She had a good job. She had a boyfriend. She was happy. She was doing just fine without God. That came to an earth-shattering stop on September 11, 2001.

At 8:46 a.m., on the morning of September 11, 2001, American Airlines Flight 11 crashed into the North Tower of the World Trade Center. In her book, *Angel in the Rubble: the Miraculous Rescue of 9/11's Last Survivor,* Genelle (now Guzman-McMillan) tells her tale of God's intervention on that terrible day. According to Guzman, she had stopped on the thirteenth floor to take off her high-heel shoes. That's when the unthinkable happened. At 10:28:22, the clock stopped for Genelle Guzman. That's when the final chapter of the first attack on United States' soil in the new millennium played its ugly hand. Just a mere 102 minutes is all it took for nineteen Islamic terrorists to murder three thousand of our fellow citizens.

Despite being hit first, the North Tower was able to stand a full twenty-nine minutes longer than her fellow (South) Tower. At 10:28 a.m. of September 11, 2001, the North Tower (WTC1) collapsed, burying Genelle Guzman with it. Her long and arduous ordeal, of twenty-seven hours, was just starting. Ms. Guzman's journey to fame was only beginning, for she would be the last survivor of the World Trade Center.

Buried under tons of pulverized concrete and mangled steel, Guzman knew that her only chance of survival was the God that she had rejected so many years before. So she prayed and prayed and prayed. For hours upon end, Genelle Guzman prayed for divine intervention. She asked for our merciful Lord to forgive her sins and give her another chance. She promised that if given another chance, she would "do his will."

Guzman was completely buried except for her left hand. Pleading with God, she reached into an open space and asked our Lord for a sign that he had heard her prayers.

According to Guzman, someone grabbed her hand and said, "Genelle, I've got you. My name is Paul." Within minutes of Paul's appearance, she could hear rescue workers. But when she was pulled from the rubble, her rescuers told her that there was nobody by the name of Paul. Guzman is convinced that Paul was an angel sent by God.

When she was finally freed, Genelle Guzman knew she was a different person. She praised God. She gave her life to God. But her ordeal wasn't over. Genelle spent many weeks in the hospital and had four major surgeries to repair her broken body. When she left the hospital, however, it was imperative to Genelle that she was baptized. She had promised God that she would get baptized. So, on November 7, not even two months after her horrifying ordeal in the rubble of the World Trade Center, Genelle Guzman was not only baptized, as she had promised God, but she also got married to her boyfriend.

Genelle made her promise to God, and she kept it. Genelle believes that God brought her through that terrible day for a reason. In an interview, Guzman said, "I think I'm here for a bigger reason and bigger purpose. My life today is a blessing. I [want] people to know about my experience, what I've been through, and how I've overcome that adversity in my life." She added, "I want people to know that God is real…that prayer works."

Chapter 15
Where Was God on September 11?
Rick Rescorla

As a young boy, Rick Rescorla grew up in Hayle, a small town in the extreme southwest corner of England. Members of the United States Army's 29th Infantry Division were stationed there while training for the D-Day invasion of Normandy on June 6, 1944. It was as a result of his contact with these men that the future director of security for Morgan Stanley/Dean Witter in the World Trade Center first became enamored with the idea of becoming a United States soldier.

In 1956, Rescorla joined the British military, first as a paratrooper and finally as a "paramilitary police inspector" in northern Rhodesia, what is now part of the central African nation of Zimbabwe. It was during this tenure that Rick met and forged a life-long friendship with Daniel J. (Dan) Hill, an American soldier. An ardent anti-communist, the young British soldier moved to the United States and joined the army. After basic training, he went to Officer Candidate School at Fort Benning, Georgia. As a newly graduated lieutenant, Rescorla was assigned to the 2nd Battalion, 7th Cavalry Regiment, 1st Cavalry Division (Airmobile).

During the Battle of Ia Drang in November 1965, Rescorla was called upon, not once but twice, to reinforce weary, battle-worn, besieged troops. The first came, as the sun was going down, on November 14. After a hard day's battle between Lieutenant Colonel Hal Moore's 1st Battalion, 7th Cavalry, and superior enemy forces under the North Vietnam communist regime, First Lieutenant Rescorla, then a platoon leader under Captain Myron Diduryk's

Bravo Company, 2nd Battalion, along with other elements of Bravo Company, landed at LZ (Landing Zone) X-Ray to reinforce the Americans for a long night's stand against the overwhelming forces.

After three days of intense combat, the American forces were able to beat back the communist forces. Rescorla's Bravo Company was airlifted out of LZ X-Ray along with the 1st Battalion. The rest of the 2nd Battalion, 7th Cavalry, however, had to march through dense jungle to reach another landing zone. During this march to LZ Albany, the 2nd Battalion, 7th Cavalry, was ambushed by superior enemy forces, just shy of LZ Albany. The fight started at 13:15. It would last for sixteen hours. The fighting was horrific.

In a jungle where they often couldn't see the enemy, the Americans were cut off and isolated in small pockets. Almost three-fourths of the battalion became casualties. Reinforcements were needed, and the men of Bravo Company, 2nd Battalion, 7th Cavalry (Rick Rescorla's outfit), despite being worn down from their own little war at X-Ray, went back into the fray. Arriving at LZ Albany just as the last rays of sun were disappearing over the horizon (about 6:45 p.m.), Rescorla and his men jumped from about ten feet off the ground (the LZ was considered too hot to land) and fought their way to the besieged soldiers.

By morning it was all over.

Thirty-six years later, sixty-two-year-old Rick Rescorla found himself, once again, right in the middle of a war zone. This time it was in New York City at the World Trade Center.

After his service, in Vietnam, Rick went to college and earned a Bachelor of Arts and Master of Arts degree from the University of Oklahoma, as well as a law degree from the Oklahoma City University School of Law.

After bouncing around in the world of academia for a while, Rick joined Dean Witter Reynolds, as they were known at that time, in 1985. Their offices were in the World Trade Center. By September 11, 2001, Rescorla was the vice president for Corporate Security at Morgan Stanley Dean Witter.

On December 21, 1988, Pan Am Flight 103 was blown up by a terrorist bomb while flying over Lockerbie, Scotland. A total of 270 people were killed, either in the air or on the ground. After this bombing, Rick Rescorla began to worry—some might say obsess—over a possible attack on the World Trade Center. Rescorla asked his army buddy, Dan Hill, to check out the security measures of the World Trade Center.

After a thorough examination, Hill, who had a background in counterterrorism, said that if he were a terrorist he would drive a truck loaded with explosives and park it in the underground garage. At the time, there was no security down there. Despite alerting the Port Authority of New York and New Jersey, which owns the site, nothing was ever done. Rick's fears proved founded, when a truck, carrying well over one thousand pounds of explosives, was parked in the garage below the North Tower (WTC1) of the World Trade Center and detonated at 12:17 p.m. on February 26, 1993.

Ramzi Yousef, the man convicted for masterminding the bombing, planned to topple the North Tower into the South Tower, thereby bringing down both buildings and killing thousands of people. His plan failed, but six people were killed, and over a thousand were injured. Well over a quarter of a billion dollars in damage were done, and some listings claim that figure at half a billion or $500,000,000. And he didn't even bring down one tower. The 9/11 attacks eight years later toppled both of the large towers as well as another one (WTC5). The World Trade Center was reopened within a month of Yousef's attack.

After the 1993 attack on the World Trade Center, Rescorla brought Dan Hill back to New York. Rescorla, still, believed that the World Trade Center was in danger and even recommended that Morgan Stanley move out of the twin towers. Hill said the next attack would be by air.

Dean Witter, Rescorla's employer, merged with Morgan Stanley in 1997 and eventually grew to be the largest tenant of the World Trade Center with over twenty floors between the forty-fourth and seventy-fourth floors of WTC1 (the South Tower) and 2,700 employees. Rescorla's office was on the forty-fourth floor of that building. They, also, had space in WTC5 with an additional one thousand employees. After the fiasco of 1993, he lost confidence in the building's management and their ability to provide adequate help in the event of another emergency. So, Rescorla started his own evacuation plan.

Over the next eight years, Rescorla changed the mind-set of the Morgan Stanley employees. He began a series of emergency evacuation practices and made these quarterly drills mandatory for everyone, including the top executives. Rick even used a stopwatch to time the employees. And anyone that was "lagging," well let's just say he let them know it.

As a result of the bombing in 1993, people started to listen to him. They may not have wanted to leave, but when Rescorla's security called a fire drill, everyone complied. It became second nature. You left the building. You didn't come back. And, by September of 2001, everyone was ready, even those who complained the most about having their day interrupted.

Rick Rescorla was in his forty-fourth floor office when American Airlines Flight 11 crashed into the North Tower at 8:46 a.m. When he heard the explosion and saw the flames, Rick knew his worst fears had come true.

Even though the Port Authority were telling people to stay in their offices, Rescorla picked up a bullhorn, which he always used in his emergency drills, and started telling the 2,687 Morgan Stanley employees, in the South Tower, to ignore the public address announcement and leave the building by way of the stairs. He also told their 1,000 plus employees in the WTC5 building to leave.

At 9:03 a.m., seventeen minutes after the North Tower was struck, United Airlines Flight 175 crashed into the South Tower. Rick Rescorla had fifty-six minutes left to live. He had a mere fifty-six minutes, and yet in that time, Rick Rescorla saved 2,681 lives. All but six of the Morgan Stanley employees, in the South Tower, survived. Rescorla and two of his security staff, along with three others perished when the tower collapsed at 9:59 a.m. Rescorla was still searching for Morgan Stanley employees when he died.

The rest of the Morgan Stanley/Dean Witter employees survived, because, just like Rick Rescorla trained them, they left the building. They did not come back. And that's why they survived that terrible day of carnage. You see, God does have a reason for everything. Sometimes we just can't see it.

Chapter 16
God of the Angel Armies

9/11 was our generation's wake-up call. It was a call to arms, still is, and probably will be until Jesus Christ returns. Those events prove there are evil people in this world and we need God more than ever. And yet, there are still many people in the world who do not recognize this fact. As I see it, we are indeed near the end times, and we need to seek God. Nevertheless, many people don't believe in God, or just don't want to.

To quote the famed Civil War general Stonewall Jackson, "He who does not see the hand of God in this is blind, sir, blind." I couldn't agree more. If two massive towers falling to the ground don't bring you to your knees to call out to the God of the universe, I don't know what will.

My time in the military taught me many things about obedience, discipline, and how to march in a straight line. But what I remember most is how God moved in my life and in my friends' lives. I have seen the other side of tragedy and what it can do to bring people to their senses, and I've seen heroes suffer for their courage. In fact, through one soldier who paid a high price, well, they ended up making a film about his ordeal.

TWA Flight 847

The 1986 movie *The Delta Force* starring Chuck Norris and Lee Marvin was based on a true incident which happened the year before. On June 14, 1985, while I was going through basic training at Fort McClellan, Alabama, two terrorists hijacked TWA Flight 847 shortly after takeoff from Athens, Greece. Over the next three days, the plane was forced to fly from Athens to Beirut, Lebanon, and on to Algiers,

Algeria, before flying back to Beirut. Some of the passengers and crew members were released, but forty of them were taken off the plane and held hostage for over two weeks. One passenger, a U.S. Navy diver, Robert Stethem, was beaten and murdered. The terrorists then threw his body off the plane and onto the tarmac at Beirut. One of the guys in my Battalion was a childhood friend of Stethem.

According to the movie, the twenty-three-year-old sailor was traveling on his U.S. Military ID card, something that we could all do. When I received my first orientation upon arriving in Germany, the MP specifically told us that we could travel anywhere in the world on our Military ID card. He recommended traveling, however, with our passports. If Stethem had been carrying a passport, the terrorists may have known he was an American but not an American service member. It possibly could have saved his life. But he was in a very, very difficult situation nonetheless.

We talked about that after the movie *The Delta Force* came out. You better believe there was a resounding standing ovation when Chuck Norris fired that rocket and blew up the bad guy in the climactic moment from the movie.

During my three years in Germany, I never saw terrorism firsthand. Bamberg, where I was stationed, was never a front for terrorist activity. My friend, Susan, who was stationed in Frankfort, had it a little differently. The first time I visited Susan and her husband, we decided to take the subway across town one afternoon. I noticed Susan made it a point to stand with her back to the subway tracks. When I asked her why she did this, she said the Muslim terrorists were known for pushing GIs in front of approaching trains if the GIs had their back to them. They all learned very quickly to have their back to the tracks, so they could scan the crowd.

In Bamberg, we never thought about it. Our main concern was if the Russians came storming through the Fulda Gap.

Germany

In 1988 I was serving in the United States Army and stationed in Bamberg, West Germany. While there, my friend, Carl—not his real name—was having a tough time with his marriage. His wife and his small daughter had decided to move back to the United States and bought tickets to fly back in late December. They were scheduled to fly on Pan Am Flight 103. The date was December 21, 1988, a date that I will remember, forever.

If you were alive back then, that date and flight number should hold significance for you. Radical Muslim terrorists put a bomb on that plane, and shortly after 7:00 p.m., it detonated. All 243 passengers, as well as the sixteen crew members, were killed. In addition, eleven residents of the town of Lockerbie, Scotland, also lost their lives. Of the 270 persons that died on that fateful day, 189 of them were from the United States. This was the deadliest foreign terrorist attack on America before 9/11.

The day they left, I went with Carl to the airport. Afterward, Carl moved back into the barracks with me and became my roommate. He received a call late that night. When he came back to the room, he was white as a ghost. I have never seen so much fear on the face of a human being.

He kept mumbling, "They're gone, they're gone."

"Who's gone?" I asked.

"My family, they're gone."

"Well of course they're gone, Carl."

"No, you don't understand. They're gone." His voice was so full of anguish. I could not comprehend what he was talking about.

"Yes, I know, they are going home," I replied.

His next words would change my life forever and allowed me to understand the fight that we—and I mean every decent human being on this planet—are in against terrorism.

"They're plane crashed."

It was as if I had been hit in the pit of my stomach by a giant welding a sledgehammer. I couldn't speak for several seconds. I saw the fear in his face. Then my reasoning came back, and all I could think of was to try to comfort my friend.

"Now, Carl, you have to understand, people survive these things all the time. You have to keep the faith and believe that they were the ones that survived this time."

I said those words, and I meant them, but what I was really thinking was, *They're gone. His family is dead. How do you deal with that?*

But I knew that he needed encouragement, so I kept trying to reinforce the idea that they had survived a crash. Of course, I had no way of knowing that it was a bomb, at over 30,000 feet, and there would be no survivors. I just kept trying to offer words of encouragement so that he could keep it together.

You can only imagine the chaos and confusion on the ground. Carl had received the call from his father-in-law. He could not get hold of anyone with the airline that could give him any viable information, so all night long; he could only dwell on the idea that his family was dead.

Finally, after what seemed like an eternity, "K" received word that his wife and daughter were alive. They had missed their flight and so were on another plane. When she landed in Washington D.C. and found out about the bombing, she had called her parents.

Their plane, from Nuremburg, to Frankfurt, West Germany, was fifteen minutes late, and they missed their connecting flight. They were not on Pan Am Flight 103 when it blew up over that small village in Scotland on that terrible day.

Fifteen minutes. Fifteen minutes of fate, or God's intervention if you will believe in that. I like to call it the grace of God.

"You just received an early Christmas present," I said.

And it's a present that keeps on giving. Those fifteen minutes are why Carl and his wife are still together today and why they are grandparents.

God has a reason for everything. We don't always see what that reason is until much later, or maybe never. But that's okay, because I know that God has the big picture.

Many years ago, my friend, Karen—not her real name—called me late one night. She was afraid that a friend of hers was going to commit suicide. She had already talked to her for a long time but was afraid that she wasn't getting through. Karen thought that maybe if I called, I would have better luck.

I didn't really want to make that call. I did not know this person all that well. She was Karen's friend, not mine. But I called, because

Karen asked me too. I talked to her for a couple of hours. Her husband was leaving her, and she had two small children. We talked and talked and talked. I used every bit of reasoning and persuasion I could.

I plainly asked her, "Do you really want *this guy* to raise your children?"

Finally, I said goodbye and hung up. I called and told Karen I had just finished talking to her friend and did not get anywhere. I told Karen that I thought there wasn't anything any of us could do to stop her. I went to bed and slept fitfully but finally did go to sleep.

The next day, I found out about just how great and far-reaching God's power is.

Karen had called her friend again, and they talked well into the night. When Karen hung up, she had the same thought as I did—it was probably the last time she would talk to her.

Then, out of the blue, she received a different phone call, this time from a friend from her past. This person said she had woken up in the middle of the night with an overwhelming urge to call her old friend. It was that conversation that convinced our friend not to take her life. Crisis averted. She never thought out killing herself again.

Why? Because God woke up a close friend in the middle of the night to reach out to her friend and share his love for her. *That is what God does.*

Chapter 17
Inspiration at the Movies From 1940 to 1976

In 2006, the American Film Institute came out with its *100 Years... 100 Cheers: America's Most Inspiring Movies*, which, as the title implies was a list of the most inspiring movies as determined by vote from over 1,500 cinematography experts. As might be expected, number one on that list is one of my all-time favorites, *It's a Wonderful Life*, starring Jimmy Stewart. What might surprise you, though, is the number of inspirational sports films that made the list. Take, for example, the fourth movie on the list was *Rocky* starring Sylvester Stallone. Not only that, thirteenth went to *Hoosiers*, twenty-second is *The Pride of the Yankees*, twenty-eighth is *Field of Dreams*, and then *Rudy* sits in the fifty-fourth slot.

According to the criteria laid out by the AFI, these movies had to:

> *inspire with characters of vision and conviction who face adversity and often make a personal sacrifice for the greater good. Whether these films end happily or not, they are ultimately triumphant—both filling audiences with hope and empowering them with the spirit of human potential.*

Granted, there are a lot of different lists out there, but I was amazed to see so many sports films hit the list. And it wasn't just the list from AFI, for example, a list compiled by a "*joe2154*" for the Internet Movie Database (imdb.com) entitled "*Best Inspirational Sports Movies*". According to this person, it was for:

> *Great inspirational movies about athletes and sports that I (he) can watch with my (his) children to inspire them and help give them vision for their lives.*

His top eight selections are right on, maybe not in the same order that I would choose, but still very good. What I can't understand, though, is how he could include a movie like *Bend it Like Beckham* as his second overall pick and yet not even rank *Rocky* anywhere on his list of thirty-two. However, I can easily go with his number one as *Cinderella Man*, with *Rudy* at number three and *Facing the Giants* at number four. *Hoosiers*, *Miracle*, and *The Natural* come in at five, six, and eight, with *We Are Marshall* falling in at the number seven spot. As you can, see many of these movies were filmed in this millennium and only *The Natural* goes back as far as 1984, so this guy is probably of our generation.

And that's what I want to look at, the top inspirational sports films that have moved me—and probably you too—over the years.

Since we are discussing some of the greatest movies of all time, let's go way back for a minute to set the stage. Let's look back at a point in history when the world was not doing so well. In October of 1940, the second war to end all wars, World War II, had been raging in Europe for over a year. Much of mainland Europe had fallen before the mighty armies of Adolph Hitler's Third Reich, and the Blitzkrieg was in full force over England. On the other side of the globe, another war between China and the Japanese Empire had been raging since 1937.

It was against this background that a new film was released in America, a movie about the legendary Notre Dame football coach of the 1920s—Knute Rockne. *Knute Rockne All American*, starring Pat O'Brien, also featured a young, up-and-coming actor named Ronald Reagan who would go on to become the fortieth president of the United States.

Then, in 1942, coming off his Academy Award-winning performance of Sergeant Alvin York in the movie *Sergeant York* (1941), Gary Cooper starred as the great baseball player Lou Gehrig in *The Pride of the Yankees*. Once again, he was nominated for an Academy Award for Best Actor, a feat that he would do again—for a third straight year—in 1943 with the film *For Whom the Bell Tolls*. Although he did not win the award, it would not be until 1953 that Cooper would actually win his second Oscar for *High Noon*, he is still most remembered for his portrayal of Lou Gehrig's famous final speech, "Luckiest Man on the Face of the Earth."

Fast forward some thirty-four years and out comes a movie written by a relatively unknown Hollywood actor by the name of Sylvester Stallone. *Rocky* first hit theaters in December of 1976 and quickly became the most popular film of the year. A simple story about a boxer from Philadelphia getting the chance to fight for the heavyweight boxing championship on the first day of the bicentennial in 1976. Rocky Balboa—a character created by Stallone after he watched the boxing match between Mohammad Ali and Chuck Wepner in 1975—was given the underdog role in a fight with the reigning, flamboyant heavyweight champion, Apollo Creed. Based on the real-life boxer Mohammad Ali, Creed could trash talk just as well as the legend himself. Talk about a David meets Goliath story, this was a true underdog story, a has-been, nobody, versus the undefeated heavyweight champion of the world.

Stallone, who was born in New York's "Hell's Kitchen" district, created a struggling character much like his own life. Perhaps that is why it resonated so much with the public. *Rocky* was nominated for ten Academy Awards and won three, including Best Picture. Four of the main actors were nominated for Best Actor awards: Stallone for Best Actor, Talia Shire for Best Actress, and both Burgess Meredith and Burt Young for Best Supporting Actor.

It is rumored Stallone wrote the screenplay in just three and a half days, and although United Artists wanted an established star, such as Robert Redford or Burt Reynolds, among others, to play the lead role, Stallone was able to convince them to let him play Rocky Balboa. It took just four weeks to film the movie and barely 1 million dollars. *Rocky* brought home $225 million at the box office, making it the highest-grossing movie of the year. Those are simply the initial numbers. Since the initial release, the *Rocky* franchise continues to sell to this day in television movie marathons, DVD sales, digital download sales, merchandise, and the list goes on. Like Coca-Cola or Walmart, Rocky has become a preeminent brand in the American psyche. Not bad for a kid from Hell's Kitchen.

Famed composer Bill Conti got his big break when he was hired to compose the musical score for this film. Until then he was an unknown. His tune "Gonna Fly Now" was the iconic sound that we all remember from that year. That now famous scene where Rocky runs through the streets of Philadelphia and then up the steps of the Philadelphia Museum of Art with that song blaring away pumped it to the number one spot on *Billboard* magazine's Hot 100 during 1977.

In 2006, *Rocky* was inducted into the National Film Registry for preservation in the Library of Congress, and in 2008, it was named the second-best sports film, of all time, by the American Film Institute.

Ever since I saw Rocky for the first time, I was hooked. Like so many other fans of the film, I received a lot of inspiration from this underdog who simply wanted to go the distance. Sometimes all the inspiration we need is to see someone trying to do their best no matter the odds. I have found several other films that have had a similar effect upon me, and I want to share them with you in the pages that follow. I have divided them up by decade for the sake of simplicity. I hope you enjoy them as much as I have.

Chapter 18
Inspiration at the Movies
The 1980s

The Natural

In *The Natural*, Robert Redford plays the thirty-five-year-old professional baseball player named Roy Hobbs who, after many setbacks, finally makes it to the Major Leagues. How important was this film? Well considering there never was a Major League Baseball team named the New York Knights and you can buy a replica of a New York Knights jersey with Roy Hobbs's number 9 on the back, I'd say it struck a chord.

The Natural, which was released in 1984 by TriStar Pictures, was based on the 1952 novel by Bernard Malamud.

In the movie, as well as the novel, Roy Hobbs, an up-and-coming baseball player, was shot and almost killed in what would have been a murder/suicide by a deranged woman named Harriet Bird. He eventually becomes a down-and-out middle-aged man trying to fulfill his dream of becoming a Major League Baseball player. Hobbs finally gets his chance at age thirty with the New York Knights as a "rookie."

By 1984, Robert Redford was one of the biggest names in Hollywood. He had become a big star, in 1969, when he teamed with Paul Newman to make *Butch Cassidy and the Sundance Kid*. He followed that with hit movies *Jeremiah Johnson* in 1972 and another Paul Newman movie, *The Sting*, in 1973, for which he was nominated for an Academy Award for Best Actor. In 1980, Redford finally won an Academy Award, although it was for Best Director, for the film *Ordinary People*. In 1984, Redford was back on the silver screen in *The Natural.*

Unlike the novel, which did not have a fairy-tale ending, *The Natural* had a happy ending when Roy Hobbs hits a towering home run to win the National League pennant. At the end of the movie, we see Hobbs playing catch with the sixteen-year-old son that he had never known about until shortly before he hit his memorable round-tripper. Iris, his childhood sweetheart and mother of the boy, is looking on. One of the difficult decisions Hobbs faces during the movie was not to accept a bribe to throw the game.

The symbolism and mythological imagery used in *The Natural* remain unsurpassed by any movie that was not explicitly based on classical mythology or a specific religious tradition. The film was critically acclaimed and was nominated for four Academy Awards.

Hoosiers

Hoosiers, a 1986 film, although fictional, is partly based on the Milan Indians, a small-town high school in Indiana. In 1954, the Indians won the Indiana State Championship of high school basketball. The Indians defeated Muncie Central High School, winners of four previous state championships including back-to-back titles in 1951 and 1952, on a last second shot by Bobby Plump. Milan had an enrollment of just 161. In 1954, there was only one class, of basketball, in Indiana. It did not matter what size the enrollment was; all high schools competed for the same title. Milan, in winning the title, was the smallest school to ever win the basketball title in Indiana.

The film *Hoosiers* is based on this miraculous team. Academy Award winner (*The French Connection*, 1971), Gene Hackman, portrays coach Norman Dale, a coach much like the real coach Marvin Wood, of Milan High. He preferred patient ballhandling over a more offensive run-and-gun style advocated by the previous coach. In those days, before the "shot clock," he would use a four-corner offense to milk time and preserve victories. In a true David vs. Goliath matchup, Hickman's Hickory High prevailed over the mighty South

Bend Central High School Bears for the state championship. They scored the winning basket on a last-second shot by their star Jimmy Chitwood. In real life it was the abovementioned Bobby Plump who took the last shot. In real life, the Milan Indians did not play South Bend in 1954. Although they had lost to the Bears in the 1953 semi-finals, in 1954, it was the Muncie Central High School Bearcats that fell to Milan in the title game.

As previously mentioned, *Hoosiers* was ranked number thirteen on the American Film Institute's Most Inspiring Movies list. The movie was, also, selected by the Library of Congress, for preservation into the United States National Film Registry.

Field of Dreams

Two years before Kevin Costner would win Academy Awards for Best Picture and Best Director, for the film *Dances with Wolves*, he starred in a movie called *Field of Dreams*. In 1987, Costner, an up-and-coming star, had played alongside Academy Award winner Sean Connery in the police drama, *The Untouchables*, a story about Elliot Ness, the FBI agent who brought down mob boss Al Capone. He followed that, in 1988, with the baseball romantic comedy *Bull Durham*, in which he played an aging minor league catcher trying to school a talented but young pitcher. One year later Costner appeared in another baseball film, but this time, instead of playing baseball, he was building his own ballpark in the middle of an Iowa cornfield. It was this surreal, mystical, and supernaturally charged movie that made Kevin Costner a big-time star.

In the 1980s, Ray Kinsella, Costner's character from *Field of Dreams*, was a struggling Iowa farmer who built a baseball diamond in the middle of his cornfield. He did so because he heard a voice that said, "If you build it, he will come."

Once the baseball field had been finished, Ray saw a young man walk out of the corn field (this was just at the edge of the outfield) and walk around in what would be "left field." Ray recognized him, from photographs that his father had, as being the famed Shoeless Joe Jackson of the infamous Chicago Black Sox, from the 1919 team that was banned from Major League Baseball after being accused of throwing the World Series. Ray Liotta, who played Jackson, in the film, is even dressed in a White Sox uniform of that era. After Ray hit some fly balls to Jackson, the figure (ghost if you wish) walked back into the corn and disappears.

The other seven members of the 1919 White Sox, which were banned from baseball, appear the next day. After that, even more players show up, this time from other teams. Liotta/Jackson even jokes that Ty Cobb had wanted to come but that they wouldn't let him, because nobody, even back then, liked the guy.

Field of Dreams was loosely based on the W. P. Kinsella novel *Shoeless Joe*. Kevin Costner and Amy Madison played the farmer and his wife, Ray Liotta was the star player, James Earl Jones portrayed Terrance Mann, a reclusive author that was patterned after J. D. Salinger, and in his final appearance, Burt Lancaster was cast as the elder Doc Archibald "Moonlight" Graham.

After he heard another voice say, "Ease his pain," and even though he is on the verge of losing his farm, to the bank, Ray was compelled to travel to Boston to find the secluded writer, Terrance Mann. After seeing a Boston Red Sox game, they soon found themselves on the way to Minnesota, where Ray has a close encounter with an aging Doc Graham on a foggy street. Ready to give up, the two decide to head back to Iowa, but on the road, they pick up a young hitchhiker by the name of Archie Graham, the young version of old Doc Graham.

The trio arrives back at Ray's home, in Iowa, to find even more players and an umpire playing a ball game. The young Graham is immediately invited to play. Ray's brother-in-law, who works for the bank, tells him that he must sell. Terrance Mann, who until then had not really believed Ray, could see these legendary Hall of Fame performers playing baseball. But the young banker, a nonbeliever, could not.

After noticing this exchange, Mann turns to Ray and says, *"Ray, people will come, Ray. They'll come to Iowa for reasons they can't even fathom. They'll turn up your driveway not knowing for sure why they're doing it. They'll arrive at your door as innocent as children, longing for* *the past. Of course, we won't mind if you look around, you'll say. It's only $20 per person. They'll pass over the money without even thinking about it: for it is money they have and peace they lack. And they'll walk out to the bleachers. Sit in shirtsleeves on a perfect afternoon. They'll find they have reserved seats somewhere along one of the baselines, where they sat when they were children and cheered their heroes. And they'll watch the game and it'll be as if they dipped themselves in magic waters. The memories will be so thick they'll have to brush them away from their faces. People will come, Ray. The one constant through all the years, Ray, has been baseball. America has rolled by like an army of steamrollers. It has been erased like a blackboard, rebuilt and erased again. But baseball has marked the time. This field, this game: it's a part of our past, Ray. It reminds of us of all that once was good and it could be again. Oh… people will come, Ray. People will most definitely come."*

After being told that he would lose his farm unless he sold it to the bank, Ray witnessed a miracle when his young daughter fell off the bleachers and started choking to death. Archie Graham hesitates only briefly before he stepped off the baseball diamond, instantly changing into "Old Doc Graham," and saved the small girl. Everything changed with that turn of events. Ray's brother-in-law can, now, finally see the ballplayers. He even asks, "Where'd they come from?"

Doc Graham, no longer able to rejoin the game, walks into the corn and disappears. The players then invite Terrance Mann to join them on the "other side." Shoeless Joe grins at Ray before he disappears into the cornfield and says, "If you build it, he will come."

When Ray looks back at the infield, he sees one player left, the catcher. And, when he takes off his catcher's mask, Ray is shocked to see his father as a young man. He finally realizes the meaning of the words, "if you build it, he will come" and "ease his pain."

As Ray Kinsella walks up to his father, the catcher asks, "Is this heaven?"

"It's Iowa," Ray replies.

"Iowa? I could have sworn this was heaven," the ghostly figure says.

"Is there a heaven?"

"Oh yeah," John says to his son, "It's the place where dreams come true."

Seeing his wife and daughter on the front porch, Ray says, "Maybe this is heaven."

As the credits role, we see the son having a game of catch with the father he never knew and a line of vehicle headlights stretching far into the night, making a pathway to the shining beacon, known as A Field of Dreams.

Chapter 19
Inspiration at the Movies
The 1990s

Rudy is a movie about a young man with a dream of playing football for the University of Notre Dame. There are so many things I like about this movie, but there are two critical moments in his journey that stand out to me.

At one point, in the movie, we find a desperate Rudy sitting in church, praying for help, when Father Cavanaugh walks up. This priest had been instrumental in getting Rudy enrolled in nearby Holy Cross University. He asks Rudy why he is there.

"I'm desperate. If I don't get in next semester, it's over and done. Notre Dame doesn't take senior transfers."

Then, a little later, Rudy tells the priest, "Maybe I haven't prayed enough."

With a snort, Father Cavanaugh replies, "I'm sure that's not the problem. Praying is something we do in our time. The answers come in God's time."

"Have I done everything that I can? Can you help me?" says Rudy.

With this, Father Cavanaugh replies, "Son, in thirty-five years of religious studies, I've come up with only two hard incontrovertible facts: There is a God, and I'm not him."

Early in the movie, the young Rudy is befriended by the groundskeeper, a man named Fortune. He was given a job by his newly found friend and even a place to sleep. But at one point, a dejected Rudy quits the team. Fortune sees him standing at one of the entrances to the Notre Dame stands. When Fortune asks what he is doing, Rudy tells him he quit because he didn't see any purpose in it anymore. Rudy says, "I wanted to run out of that tunnel, for my dad, to prove to everybody that I was—"

"Prove what?" Fortune asks.

"That I was somebody."

"Ah, you're so full of crap. You're five-foot nothin', 100 hundred and nothin', and you have barely a speck of athletic ability. And you hung in there with the best college football players in the land for two years. And you're gonna walk outta here with a degree from the University of Notre Dame. In this life, you don't have to prove nothin' to nobody but yourself. And after what you've gone through, if you haven't done that by now, it ain't gonna never happen. Now go on back."

Rudy returns to the team and goes back to practice. He soon discovers his name is not on the roster for the final game of the season. The movie then shows a scene where all the seniors place their jerseys, one by one, on the desk of head coach Dan Devine. They each tell him they want Rudy to play in their place. A powerful scene.

Rudy is then allowed to suit up for the last home game on November 8, 1975. After watching the whole game from the bench, Rudy is summoned on the field for the final three plays of the game. In the last three plays of the game, Rudy participates in a kickoff, an incomplete pass, and on the very last play, Daniel E. (Rudy) Ruettiger records a sack. The crowd goes wild. Notre Dame defeats Georgia Tech by a 24–3 score and Rudy is carried off the field on the shoulders of his teammates.

It was the first time in the long and storied career of Notre Dame that a player was carried off the field. Not even the famed Four Horsemen or "The Gipper" were accorded this honor. Only one other player has been carried off the field, and that was twenty years later when Mark Edwards, the fullback on the 1995 team, ran for three touchdowns, a two-point conversion, and even threw for a two-point conversion while leading the Fighting Irish's upset of the fifth-ranked USC Trojans on October 21, 1995.

One point of interest: the movie gives you the impression that this was the last chance that Rudy would have to play for the Fighting

Irish. That wasn't so. Although it was their last home game of the season, they still had to play two games on the road.

For the Love of the Game

Michael Shaara, who won the Pulitzer Prize in 1975, for his award-winning historical novel *The Killer Angels*, a story about the Battle of Gettysburg, put to pen what would become his final work, the baseball novel *For Love of the Game*. Unfortunately, he passed away, in 1988, before he could see it published. His son, Jeffrey Shaara, who would go on to finish the trilogy of Civil War works, found this delightful novel about baseball and in 1991 had it published. Kevin Costner then brought this story to the silver screen in 1999.

For Love of the Game is the story of an aging Major League Baseball pitcher played by Costner facing questions about life after baseball. Billy Chapel is about to be traded. He is forty years old, his throwing arm was shot, he's had the worst season of his career, and he finds out the Detroit Tigers have been sold and the new owners want to trade him after the season is over. It's the next to last game of the season, and on top of that, the woman he loves will be leaving forever, to go to London, England.

His team was playing the mighty New York Yankees in the famed "House that Ruth Built" otherwise known as Yankee Stadium. Billy was slated to pitch the next to last game of the season, and his manager wanted to bench his catcher and friend, Gus Sinski. His manager wants a "hitter"; Billy wants his friend to catch. Billy prevails. In a low-scoring affair with only one run scored, and it was the slow-as-molasses Sinski that gave Chapel all the offense he would need. Billy Chapel prevails against the mighty Yanks.

But the real story of the film is Billy's pitching interspersed with his life with New Yorker, Jane Aubry. Throughout the movie, Billy reminisces about how he met Jane and how they managed their slowly evolving relationship until Billy dashes it by pushing away the blond beauty played by Kelly Preston after he suffers a freak injury that threatens his career. Only at the last does Billy truly realize how much he loves her and how important she is to him.

One of the more poignant moments in this movie was during the bottom of the eighth inning, when Billy looked at the scoreboard and sees nothing but zeros for the Yankees. It's at this point that Billy Chapel starts to realize just what is going on.

Gus Sinski, Billy's long-time catcher, sees that something is bothering his pitcher, so he calls time out and trots out to the mound. His pep talk to Billy was truly inspiring.

"What's the matter?" Gus asks his old friend.

"I don't know if I have anything left," Billy says as he turns around and takes a couple of steps toward second base.

"Chappie," says Gus as he looks his pitcher in the eyes, "You just throw whatever you've got. Whatever's left. The boys are all here for you. We'll back you up. We'll be there. 'Cause Billy, we don't stink right now. We're the best team in baseball, right now, right this minute because of you. You're the reason. We're not going to screw that up. We're going to be awesome for you right now… Just throw!"

Prior to the start of his final inning as a Major League pitcher, famed broadcaster Vin Scully had this to say about Billy Chapel:

> *At forty years old, Billy Chapel is flirting with perhaps the greatest accomplishment in baseball. And, standing in his way will be Matt Crane hitting for Babe Nardini, then Jesus Cabrillo and Ken Strout has a bat in his hands in the dugout and might very well get the call to bat for Hymie Ruiz. And, you know, Steve, that you get the feeling that Billy Chapel isn't pitching against lefthanders, isn't pitching against pinch hitters; he isn't pitching against the Yankees. He's pitching against time. He's pitching against the future, against age and, even when you think about his career, against ending. And, tonight, I think he might be able to use that achy-old arm one more time to put the sun back up into the sky and give us one more day of summer.*

Another great fielding play, of which there were many, followed by Chapel's ninth strikeout, brought the game down to the final out. The final batter standing in the way of pitching immortality was rookie Ken Strout, whose father had once played with Billy Chapel on the Detroit Tigers team. With the musical score rising to a crescendo and Billy's on-again-off-again love interest and her daughter, watching from opposite ends of America, one last amazing play saved the day and let Chapel finish his major league career on the highest note possible—a perfect game!

No feel-good movie would be complete without a happy ending, and *For Love of the Game* does not disappoint. Billy Chapel went to the airport, the next day, to catch a flight to London. He finds his love, Jane Aubry, also waiting on a plane for England, since she had missed her flight to watch his perfect game. When she asks why he's there, Billy tells her how he really feels, and, as the credits start to roll, we are left with the two lovestruck denizens kissing while on their knees in the airport.

The movie was not only motivating, from a human interest stand point, but it also shows how real God is and how important it is to keep everything in the proper prospective.

God does truly have all of the answers.

Chapter 20
Inspiration at the Movies
The 2000s

In the spring of 1999, coach Jim Morris of the Reagan County Owls baseball team, in Big Lake, Texas, made his team a promise that if they won the District Championship, he would try out for Major League Baseball. The Owls had never won a championship, but in a reminiscence of the old line, "win one for the Gipper," they won one for their coach.

So, at the ripe old age of thirty-five, Morris tried out for the Tampa Bay Devil Rays. Although it was considered a joke by the professionals that were there that day, jaws probably hit the ground when they watched this retirement age guy, hurl not one, not two, but twelve straight 98-mph fastballs. After playing for the Devil Rays' minor league organizations, Orlando Rays (Double-A) and Durham Bulls (Triple-A), Morris got his chance to finally pitch in the Big Leagues.

On September 18, 1999, Jim Morris was brought in as a relief pitcher during a game with the Texas Rangers in Arlington, Texas. This is the premise of the 2002 movie *The Rookie* staring Dennis Quaid as Jim Morris.

Despite a less than encouraging endorsement from his father, and with some trepidations, Morris signs a contract and starts pitching with the Tampa Bay Double-A farm team but quickly moved up to the Durham Bulls, a Triple-A (one step below the Major Leagues) team. Although doing well, as a relief pitcher (in fact his coach told Morris that he had been his best reliever that month), Morris was ready to quit and go back home. His wife, however, was able to talk him out of this decision.

Finally, his chance comes when he was called up in September. That night, with his family, and many of the townspeople from Big Lake, in attendance, Morris strikes out Royce Clayton to end an inning. His big moment had arrived, and he had wholeheartedly embraced it.

After the game, Jim's father approaches him and tells him he is proud of him and apologizes for not being there for him. His wife meets him, and they walk out of the park. We then see a large crowd of people from his home town waiting for him.

At the end of the movie, you see a trophy case in the Big Lake High School.

In the case is Morris's Tampa Bay jersey.

Miracle

Miracle, a 2004 Walt Disney Production film, is the second movie to be made about the miraculous come-from-behind victory, over the Soviet Union, by the United States Olympic Hockey Team during the 1980 Winter Olympics in Lake Placed, New York. The first one, *Miracle on Ice* (1981), came out shortly after the Olympic magic. That made-for-television movie, or should I say "docu-drama," featured Karl Malden as coach Herb Brooks along with Steve Guttenberg, who played goalie Jim Craig and Andrew Stevens as team captain Mike Eruzione. They used real game footage as well as the recorded commentary from Al Michaels.

In the 2004 film, we have Kirk Russell playing the role of Herb Brooks and, well, a bunch of hockey players, not actors. Whereas in the first movie *Miracle on Ice*, you had some Hollywood heavy hitters like Malden, Guttenberg, and Stevens, the cast of *Miracle* were mostly journeymen in the acting world. There were only a few that I recognized, and none of them were the players. Director Gavin O'Conner purposely hired hockey players to play the roles. He thought it would be easier to teach hockey players how to act than to teach actors how to play hockey. Of course, you do have the announcers reprising their role as the voice of the free world, Al Michaels with his color commentary Ken Dryden along with Jim McKay, best remembered for all those years of being the host of ABC's *Wide World of Sports.*

The 1980 Winter Olympics were hosted in Lake Placid, New York, and the world was rife with political turmoil as the Cold War powers—the USA and The Soviet Union—were frozen in a nuclear stalemate. At this point in time Afghanistan was the powder keg, and the Soviets were holding the proverbial "match."

Russia, a former name for the Soviet Union, had invaded Afghanistan in December of the previous year. President Jimmy Carter was considering a boycott of the 1980 Summer Olympics, which was scheduled to be held in Moscow. As a result, the Soviets along with the rest of their communist-bloc allies threatened to boycott the Lake Placid Olympics. And, although the Americans would boycott the Summer Olympics, held later that year, the Soviet Union led the communist-bloc nations to Lake Placid for the Winter Olympics.

Whereas the hockey team fielded by the Soviet Union in 1980 was, for all practical purposes, a professional team, the young United States team was made up of a bunch of college players, whose average age was twenty-one. Coach Herb Brooks personally recruited each player to represent the red, white, and blue in the Olympics.

The Soviets had won six of the seven previous Olympic gold medals, including the last four. Only an American upset in the 1960 Olympic Games marred that impressive string. By 1980, the red juggernaut that had become the Soviet Union hockey team was considered unbeatable. The previous year, this Soviet national team, some of which were actually in the military, had defeated the NHL all-star team by a 6–0 score. Basically, in a football game that would be the same as a 42–0 skunking. With team captain Boris Mikhailov and Vladislav Tretiak, the "best goaltender in the world" leading them, this Soviet Union team was the "heavy" favorite to make it five straight Olympic gold medals.

Facing this mighty red hoard from Russia, Herb Brooks put together a handpicked team that consisted of just one returning player from the previous (1976) Olympic team. That was Buzz Schneider.

His team had one purpose—to beat the Russians. Almost half (nine) of the twenty-man roster that Brooks settled on had played hockey for him at the University of Minnesota.

Going through a grueling sixty-one exhibition game schedule in only five months, the Americans finished with a match against the Soviets just two weeks prior to the start of the Olympics. They were thoroughly schooled by the Red Menace from Moscow, bowing to them by a 10–3 score. Brooks wanted this exhibition game against the Soviets, because he wanted his players to get it out of their system. That way they would not be in awe when they met them in the Olympics. Soviet coach Viktor Tikhonov later said that the game made his players overconfident. That was an understatement. The Russians, so used to winning easily, came to play a game of hockey. The Americans came for war.

After a wild come-from-behind 2–2 tie against Sweden, in the opening round, the young American squad stunned the communist world with a 7–3 victory over Czechoslovakia, the supposed second-best team in the Olympics. They followed with wins over Norway, Romania, and West Germany. With a 4–0–1 record, in the Group Round, the brash upstarts from our side of the pond were scheduled to play the Soviet squad, on February 22, 1980, in the medal round. Sweden and Finland played in the other game.

The Soviet team jumped out to an early lead, but the Americans tied it at 1–1 with a Buzz Schneider goal at the 14:03 mark of the first period. Down 2–1 with the final seconds of the opening stanza, Mark Johnson took a rebound and punched a last second shot past Tretiak to tie the game at the end of the first period. The Soviets scored a single goal early in the second period to set the stage for some late-game dramatics. Once again, the American team found themselves trailing the Russians, by a 3–2 score.

Vladimir Myshkin, the backup goalie for the Russians, had replaced the "best goaltender in the world," Vladislav Tretiak, after the supposed first period meltdown. Soviet coach Viktor Tikhonov, who made that decision, later said that it was the biggest mistake of his career. Although Myshkin did hold the Americans scoreless for the next twenty-eight minutes, he would give up two crucial third-period goals as the U.S. team stormed back to take a 4–3 lead with ten minutes to play.

The first goal came when Mark Johnson picked up a loose puck and fired it past Myshkin, for his second goal, eight minutes and thirty-nine seconds into the final frame of this all-important matchup between the superpowers of the world. Less than two minutes later, Mark Pavelich passed to an open, team captain, Mike Eruzione, who fired the go-ahead puck past Myshkin. This gave the U.S. team their first lead. There were exactly ten minutes left to play.

This is where the brilliance of Herb Brooks really shined. With their backs against the fall for the first time, the Soviet team came out with a vengeance. The Reds, which outshot the Americans 39–16 for the game, repeatedly fired shot after shot, only to see stellar defensive play by Brooks' boys, especially goalie Jim Craig, who had come close to losing his job before the Olympics even started. Brooks kept saying, "Play your game. Play your game." Instead of going into a NFL-style "prevent defense," the Americans kept attacking the Soviets and pressuring them into making mistakes. Their shots became more erratic as the desperate Russian team tried to fight back. With less than a minute to play, the Soviets got the puck back. There at the end, they had several shots on goal, all of which Jim Craig defended successfully. To the Americans' disbelief, Tikhonov never pulled his goalie in the final minute. He just couldn't conceive of this happening, so he never prepared for it. Finally, with the clock ticking down, Mark Johnson scrambled for the loose puck and successfully passed it to Ken Morrow. With the crowd counting in the background, the now-famous call by ABC announcer Al Michaels was heard, not live, but by tape delay from that small backwoods hamlet in New York. "Eleven seconds. You've got ten seconds, the countdown going on right now! Morrow, up to Silk. Five seconds left in the game. *Do you believe in miracles? Yes!*"

And with that, pandemonium broke out in Lake Placid, New York. The young group of upstarts from the good ole U.S. of A. had shocked the world. But, it would not be over. This was not a typical "final four"-style tournament like it is now. It was not the two winners from the first, or semifinal, matches that would play for the gold medal. By Olympic rules, at that time, the medal round was a round-robin, and since the Americans had that tie to open the "group play" with Sweden, the Soviets could still win the gold medal if they defeated Sweden while the Americans lost to Finland. In fact, the United States could have finished fourth, depending upon points scored by all four participants.

Once again, the United States hockey team found itself trailing an opponent in the 1980 Olympic Hockey games, this time by a 2–1 margin. But, in true Herb Brooks fashion, they came back and took home the gold with a 4–2 victory.

Cinderella Man

In *Cinderella Man*, Russell Crowe, the Academy Award-winning actor, played Jim Braddock, a seemingly washed-up boxer. This movie was about James J. Braddock, the Cinderella Man. In a time of abject misery for so many (the Great Depression), Braddock became the shining beacon for so many. In a land of lost, he became the symbol of hope for millions. Academy Award winner Renee Zellweger was his wife, Mae, and Paul Giamatti was nominated for an Oscar as the Best Supporting Actor, for playing Joe Gould, Braddock's manager/trainer. Zellweger had won the Best Supporting Actress Oscar in 2003 for the movie *Cold Mountain*. Russell Crowe had won his award for his sterling performance in *Gladiator* in 2000.

Early in his career, Braddock was virtually unstoppable. One source lists his record, after his first three years as being 44–2–2. However, that same source, in a rundown of all his fights, shows him having twenty-three defeats by the time that he won his forty-fourth fight. Even with the advent of the computer, along with Google and Wikipedia, it can still be very difficult to iron out the facts sometimes. What we do know is that James J. Braddock had been a great fighter before he broke his hand. After that, his career went down, and he was eventually kicked out of the sport that he loved so much. And this was happening, just as the Great Depression was setting in.

Unable to continue in his chosen profession, after he broke his hand in a fight with Abe Feldman on September 25, 1933, Braddock had to resort to getting whatever jobs he could. For a time, he worked on the docks as a longshoreman. But even then, it wasn't enough. Finally he had to turn to the government for "relief money," which we now call welfare, just to feed his family.

Back in 1928, Braddock had been on top of the game when he defeated highly touted Tuffy Griffiths in a second-round TKO. The following year, however, saw Braddock break his hand for the first time, this time in a title bout with Tommy Loughran. For the next few years, Braddock was only so-so in the ring, losing more fights than he won. After breaking his hand, again, on September 25, 1933, Braddock found himself out of boxing.

In 1934, however, Jim Braddock shocked the boxing world with his third-round knockout of "Corn" Griffin. The fight, which was supposed to be just a tune-up for the up-and-coming boxer from Blountstown, Florida, was stopped in the third round by Braddock's hard right hand. With a newfound resurgence, Braddock started the comeback trail when he defeated John Henry Lewis with a ten-round decision. Two years before, during Braddock's long term of suffering, Lewis had beaten him. This time, it was Braddock's turn. Four months later, the return was complete as Braddock pounded heavyweight contender Art Lansky into submission. The fifteen-round "unanimous" decision made him the number one contender which meant that heavyweight champion Max Baer had to give him a shot at the title.

Despite being given no chance at all, Jim Braddock, on June 13, 1935, became the Heavyweight Champion of the World when he subdued the cocky Baer in a unanimous decision. The metamorphosis from aging has been to top dog was complete. For all the downtrodden little guys, he was a national inspiration.

And of course, there was the year 2006, when four outstanding, inspirational sports movies came out.

Invincible tells the tale of rags-to-riches, thirty-year-old bartender, Vince Papale, who tried out and won a spot on the Philadelphia Eagles football team in 1976. *We are Marshall* is about the rebuilding of the University of Marshall's football program from ground-up after virtually the entire team as well as their head coach and most of his staff were killed in a tragic plane crash while coming back to Huntington, West Virginia, on November 14, 1970. *Glory Road* was a movie about the improbable run to the National Championship by the Texas Western (now University of Texas El-Paso) Miners basketball team in 1966. And of course, my movie, the movie that changed my life, *Facing the Giants*, which first aired on September 29, 2006. What makes *Facing the Giants* so good is that all of the abovementioned movies are about man and his endeavors; *Facing the Giants* is about *God.*

Invincible

Invincible is the story about one man's quest for NFL greatness. In the movie, Vince Papale, played by Mark Wahlberg, is depicted as an out-of-work high school teacher whose wife had just left him. He got a part-time job at his friend's bar in South Philadelphia, Pennsylvania, where he met his future wife, Janet, who also started working at the bar. She was a New York Giants fan, a hated rival of the local Philadelphia Eagles team. In reality, Papale's wife had left him, along with a nasty note, but that had been five years before the events portrayed in the movie. And he did not meet his future wife, Janet, until after he had quit playing pro ball. They did not get married until 1993. But you know how Hollywood likes to mess around with things like true facts.

In the movie, Papale along with hundreds of men shows up for an open tryout, held by the new head coach of the Philadelphia Eagles, Dick Vermeil. The truth was that Vince had actually played semi-pro football for the Aston Green Knights and then two seasons for the Philadelphia Bell of the now defunct World Football League. Although Papale did participate in an open tryout, that was for the Philadelphia Bell. As a result of his time with the Bell team, Papale received an invitation for a private workout with the Eagles. After that workout, he was invited to try out for the team. After earning a roster spot, Vince would spend three years with the Eagles playing mostly on the special teams, although he did catch one pass for fifteen yards as a wide receiver. An injury forced Papale to retire after the 1978 season. He played in all but three regular season games during his time with the Eagles and was voted special teams captain by his teammates in 1978. In 1980, two years after Papale retired, the Philadelphia Eagles, with many of his former teammates, won the NFC title and played in their first Super Bowl, on January 25, 1981. The game was played just five days after the Iranian hostage crises ended. On January 20, just twenty minutes before incoming president of the United States, Ronald Reagan, took office, the Ayatollah Khomeini, the in de facto dictator of Iran, released the remaining fifty-two American citizens who had been kidnapped and illegally detained since 1979 when an Iranian mob stormed the United States Embassy in Tehran. It seems that for all of his rhetoric about fighting the great Satan, us, otherwise known as the good guys, or rather the United States of America, Khomeini did not want to mess with the United States after we got a real president in office.

In the movie, Papale is depicted as scoring a winning touchdown when he recovers a fumbled punt and runs it back all the way. In truth, the Eagles won rather handily by a 20–7 score (the Giants got their lone score late in the game). The fumble did happen, but he was not allowed the score, as the NFL rules of that time did not allow a fumble to be returned. The Eagles received the ball where he recovered it and later scored a touchdown.

We Are Marshall

We Are Marshall is a film depicting the aftermath of a 1970 plane crash that gutted the Marshall University football program. Thirty-seven players, the head coach, and most of the staff were killed. It was the second such plane accident that happened to a college football team in barely over a month. The starters of the Wichita State University football team along with the coach and athletic director and others were flying in one of two Martin 4–0–4 twin-engine aircraft, on October 2, 1970, when it crashed in the Colorado Rocky Mountains near the Loveland ski area west of Denver, Colorado. Thirty-one of the forty passengers and crew were killed in this crash.

Forty-three days later, a similar catastrophe struck the community of Huntington, West Virginia. In all, thirty-seven players from the Marshall University Thundering Herd football team lost their lives, on November 14, 1970, when their chartered airplane crashed on approach to Huntington. Head coach Rick Tolley and five members of his staff, along with Marshall's athletic director, team trainer, sports information director, and even the radio play-by-play announcer Gene Morehouse, perished in the tragic crash. In all, seventy-five lives were lost, which included five crew members and twenty-five team boosters. Five doctors, half of Huntington's physicians, were among those twenty-five boosters. They were returning home from an away game at East Carolina University in Greeneville, North Carolina.

In the movie, university president Donald Dedmon, played by David Strathairn, is moved to hire Jack Lengyel, who is portrayed by Academy Award winner Matthew McConaughey, to be the new football coach. In the movie, Lengyel convinced assistant coach "Red" Dawson to return for one year to help rebuild the program. Dawson had survived the horrific night in November only because he had left on a recruiting trip immediately following their game and so was not on the ill-fated plane. Matthew Fox, playing Dawson, brought to life just how much suffering the man went through while battling what we now call "survivor's guilt" in the movie. He promised Lengyel he would come back and coach for one year. He kept that promise. After 1971, he never coached again.

For those of you who weren't around back in 1970, when you hear about Marshall University, you might think "Randy Moss," the great wide receiver of the 1990s. Maybe you remember his quarterback from those days, Chad Pennington, who, like Moss, went on to play professional football in the NFL. Or maybe you remember the 1990s when Marshall won 103 games, between 1991 and 1999, in nine years. Their .851 winning percentage was topped by only Florida State (.896) and Nebraska (.881). Twice they were undefeated (1996 and 1999), and twice, in 1992 and 1996, they won the NCAA Division 1-AA National Championship.

But for those of you who were alive in 1970 and kept up with football, especially college football, you remember the tragedy that befell the small West Virginia community of Huntington when they lost their football team. It likens to the American Civil War when many small towns lost entire generations of families as fathers and sons, and perhaps even grandsons went off to war and never returned. It's just something that you cannot plan or prepare for.

After the crash, the rest of the season was cancelled. President Dedmon even considered suspending the football program. The decision, however, was made to rebuild the program, mainly by using the remaining players, three varsity players who missed the trip due to injuries or other reasons and the members of the freshman squad, who were ineligible to play, as freshmen. That rule was graciously lifted by the NCAA for Marshall in 1971 and then for all NCAA schools in 1972. The rest were walk-ons, of which three were basketball players who used their fifth year of eligibility to play on the football team.

Joe McMullen, who had once coached under Joe Paterno, at Penn State, was hired to replace the athletic director. Dick Bestwick, an assistant coach at Georgia Tech, was hired to be the head coach, but he quit after two days. Finally, Jack Lengyel, a coach at a little-known private college in Wooster, Ohio, asked for and was hired as the Thundering Herd's new head coach. As previously stated, Lengyel was able to convince "Red" Dawson to help rebuild the team. Dawson promised he would stay for one year, and that is exactly what he did. After the 1971 season, he never coached again. Defensive back Nate Ruffin, along with defensive back Felix Wright and defensive lineman Eddie Carter, returned to help lead the youngsters, or as Jack Lengyel called them, the Young Thundering Herd, in their debut as a team that had literally arisen from the ashes. Ruffin and Wright had missed the game due to injuries, while Carter stayed at home for personal reason.

There weren't many of them, that first year, and they weren't very good, but they did play. President Richard Nixon sent them a message which said, "Friends across the land will be rooting for you, but whatever the season brings, you have already won your greatest victory by putting the 1971 varsity on the field." Coach Lengyel read that letter on the first day of practice.

They lost their first game at Morehead State but were able to regroup and win their next game, the first home game of the season, against Xavier University, on a last second touchdown pass from Reggie Oliver to fullback Terry Gardner. They only won one other game that year, but at least they played, or as the president of the United States said, "They put the 1971 varsity on the field."

Glory Road

Glory Road is based on Don Haskins' autobiography of the same name. The book, which was published in 2005, tells the story of how a "Cinderella" team, the Texas Western Miners (now the University of Texas El Paso[UTEP]), were able to overcome all obstacles, including the famed Adolph Rupp and his powerhouse, Kentucky Wildcats, to win the 1966 NCAA Basketball Championship, even though he started an all-black (African-American) team in the pivotal championship game.

The book, which was reprinted five times in the first four months, tells about Haskins' life as a player, women's basketball coach, and then about the remarkable run to an improbable National Championship. *Glory Road*, the movie, starring Josh Lucas as Coach Don Haskins, was actually the top box office film during the week of January 15, 2006, earning well over $13 million. The movie, which alludes to his time as a women's coach, really is just about the 1966 season. Although the movie gives the impression that Haskins' first team at Texas Western won the championship, in reality it was not until his fifth year in El Paso that they brought home the national title. With 719 career wins, Haskins is tied for nineteenth all-time wins, but John Calipari, with 694, will probably pass that this year.

The 1966 season started with a victory, and for the next twenty-two games they saw an additional number added to the win column. They were finally defeated by two points in their last game of the season. Still, a 23–1 record and ranked, by AP and UPI, as the third best team in the nation was nothing to sneeze at. The

Miners were the fifth seed in the Midwest Region. When you think about the (almost seventy teams) that compete in today's NCAA Tournament, it's hard to contemplate there were only twenty teams in 1966. Put into this perspective, the Miners were considered one of the best twenty teams in the country going into tournament play. They topped Oklahoma City University (modern day Oklahoma State University) rather easily by an 89–74 margin. They followed that with back-to-back barn burners, first with an overtime victory over Cincinnati and then a Regional Final double-overtime win over the University of Kansas. In the game with Kansas, famed Jayhawk star Jo Jo White, who would go on to a stellar career with the NBA Boston Celtics, appeared to have made the game-winning basket as time expired in the first overtime, only to have it overruled by an official who noticed that he had stepped on the sideline just before shooting. There has been a lot of controversy about that shot. Did he step on the line or not? Ask Jo Jo White, and he will say no. Don Haskins had this to say about the game-changing decision by referee Rudy Marich, "It was the right call." A deflated Kansas squad lost, by a single point, in the second overtime, 81–80.

In the semifinals (they didn't start calling it the Final Four until 1978), number one-ranked Kentucky rallied to beat the second-ranked Duke Blue Devils, while the third-ranked Miners toppled Utah by an 85–78 margin.

Adolph Rupp's vaunted Wildcat team, led by all-American Louis Dampier and Pat Riley, who would go on to win five NBA championships as a coach, came out full of confidence that they would throttle the upstart team from El Paso, Texas. It was the scrappy Miners, however, that dictated the tempo of the game. From center David Latten's slam dunk, something he had been doing all year, at the beginning of the game, to Bobby Joe Hill's back-to-back steals midway through the first half, to the excellent foul shooting down the stretch, the underdogs took charge and held sway throughout most of the game. In fact, after Nevil Shed's free throw, with a little over ten minutes left in the first half, gave the Miners the lead, Kentucky was never able to regain their luster. Although the Wildcats did close it to within a point early in the second half, it was not enough. Bobby Joe Hill, the fast-moving, hard-charging guard, from Detroit Michigan, led all scorers with twenty points, but perhaps none were more important than the ten-second span in the middle of the first half when he, not once, but twice, stole the ball and drove for easy layups. Even venerable Adolph Rupp would later admit those two plays made a huge difference in the game. Hill, who along with Orsten Artis and Willie Worsley, played all forty minutes, was constantly making big play after big play. Big David Latten was intimidating under the goal, but he also hit all six free throws while Artis bottomed all five of his, and Willie Cager put in all but one of his seven shots. When you consider that Kentucky took seventy shots from the field, compared to only forty-nine by the Miners, and the Cats made five more baskets, that 17–18 from the foul line made a huge difference. Take in account, how many times during the season that Texas Western was short changed by (perhaps) nearsighted officials (in the last game of the season, at Seattle, not one personal foul was called on their opponent), the Texas Western Miners outshot the Wildcats, making twenty-eight of thirty-four free throws to eleven of thirteen. Those seventeen points were the difference.

Perhaps the movie got it right, when they show Lukas/Haskins telling his team that he is only going to play the black players to show everyone that they could do it. Or perhaps it is what the real Haskins said when asked about his decision. He said that he "just wanted to put (his) five best guys on the court." In his book, Haskins said that he wasn't trying to be "some racial pioneer." The movie, however, like so many Hollywood productions, often changes the facts for reasons only known to them. We'll never know, but one thing we do know is on March 19, 1966, in a true David vs. Goliath story, a team of mostly unknown black basketball players brought down the reign of a mighty giant and revolutionized the way business was on the hardwood court, in the South.

Chapter 21
Inspiration at the Movies
The Movie that Changed My life

Even though we have reviewed the top inspirational films of the last few decades, there is only one sports film that truly changed my life: *Facing the Giants.*

Facing the Giants is a 2006 independent film made by the Kendrick Brothers through Sherwood Baptist Church in Albany, Georgia. It was the first of its kind, a movie made solely by a church for a church audience. Alex and Stephen Kendrick wanted to teach the kids in the community about God and Jesus. This movie not only changed the lives of the people in southern Georgia, it changed mine. I especially enjoyed the back side of the DVD which tells the story of how it was made, produced, and marketed. It made me see God in a different light. It especially made me want to read the Bible. Since that day in 2007, I have read the Bible multiple times and many specific sections over and over again. I now want to learn more about God, and I attribute that hunger to *Facing the Giants.*

Facing the Giants was a true independent film. It cost a little over $100,000 to make and earned well over $10 million at the box office and in DVD sales. In the movie, Coach Grant Taylor is the coach of the Shiloh Christian Academy football team, nicknamed the Eagles. Everything seems to be going against Coach Taylor. He has never had a winning season on the gridiron. He finds out that, once again, his best player is transferring to another school for his senior season. In addition to this, he and his wife just can't seem to have a baby, his automobile is always breaking down, and there is a horrible smell in his house. To make matters worse, he finds out that some of the more influential fathers of his players are trying to get him fired so that one of his assistants can be elevated to head coach.

With his confidence and self-esteem at an all-time low, his faith severely shaken and seemingly nowhere to turn, an old man walks into his office and says, "Coach Taylor, the Lord is not through with you here."

This man tells him a story, much like many of the parables that Jesus Christ would use when telling the masses that followed him. It was about two farmers that desperately needed rain for their crops. They both prayed to God, but while one of them did nothing else, the other went out and tilled his land and planted the seed. When the old man said to Coach Taylor, "Which one do you believe had more faith?" He then told the coach, "You have to prepare the field, and God will bring the rain in his time."

That has become my catchphrase. I will prepare the field, and I will trust in God to bring the rain, when he sees fit. That's good enough for me.

Shortly thereafter, Coach Taylor made a life-changing decision, and he shared it with his players. He told them that it had to be for God's glory. Playing to win just didn't do it anymore. Everything we receive is because of God's grace. He said that they would play for God. When they won, they would praise God. And when they lost, they would still praise God.

Without going into the movie too much, I do not want to give away the story, because I believe you should watch it yourself. Coach Taylor's life does turn around, and his team starts to win. They even have a small player transfer to their school. He was a former soccer player turned placekicker, named David, and the team they have to play in the championship game is called the Giants. Need I say more?

On a more personal front, Coach Taylor's life also takes on a major transformation. "I will still love you." Taylor's wife cries out as she looks toward heaven. "I will still love you." That has become my battle cry as I face life's challenges. Because I know God has my back, no matter how bleak things may look.

Facing the Giants is a great, heart-warming movie. And yet, very few people have heard about it. Buy it, rent it, or if you see it listed in the television listings, then just turn on your television and watch it.

The movie changed my life. It inspired me to want to know God better, and it motivated me to "want to" read *the Bible*. Oh, I had read parts of *the Bible* but never all of it and never with a real understanding of it. Like most people, my experience with the book was the King James Version. And I'm sure you understand just how difficult it can be to understand it.

I asked a friend who is well versed in *the Bible* for a little help. His recommendation was *God's Word*, by God's Word to the Nations Bible Society. It was published by World Publishing, Inc., of Grand Rapids, Michigan, 49418 USA., in 1995. I quickly went on line, found a place that I could purchase a copy, and then bought it. However, I soon found out that I had bought a copy of *The New Testament*, only. There is nothing wrong with that. It is about Jesus Christ and the beginning of Christianity. But it was not the entire *Bible*. After reading it, I decided to get the full version, so that I could go back to the beginning (Adam and Eve) and read *the Bible* all the way, from beginning to end.

You see, when I first bought my copy of *The New Testament*, I would just open the book and, without looking, point to a certain point and start reading from there. I might read a single page or it might be five to ten pages. You would be surprised to find out how often that would result in a personal message from God that was really relevant to my life and what was going on at that time. It was almost like God was sending me a message that I needed at that time. I still do that from time to time, but as time progressed, I did want to go back to the beginning and read it in order. Reading *God's Word* was a pleasure.

Yes, there are parts that are still hard to understand or just plain difficult to read. *The Book of Job*, for example, is just plain hard. And it is so long; however, it is well worth the reading to those of you who will persevere to the end. As I have told my friend "T" many times that she should read the *Book of Job*, I have read it more than once, although not as often as other parts of *the Bible*. And, Job, remember, despite being in such terrible pain and having lost everything, still did not give up on God. And neither did God give up on Job.

One day, I saw an advertisement on television about a new version of *the Bible*. It was called *The One Year Chronological Bible*. As it was advertised, this is a new version of *the Holy Bible, New Living Translation*, which was first published in 1996. In 2007, this second edition of the *New Living Translation* was published. When I saw the advertisement, I told my good friend, Steve Rydell, about it. And, that Christmas, of 2009, he gave me a copy as a present. The rest, as they say, is history.

This version of *the Bible* was written in chronological order, rather than by the various books, and is broken down into small, daily, sections so that a person reading anywhere from three to ten pages daily could read the entire *Bible* in a year. It usually was about four to five pages a day, and it made sense. I am a historian by nature and passion. I like to study what had happened in the past so as to not make the same mistakes in the future.

My major problem before reading *The One Year Chronological Bible* was the confusion that often came from reading the account of a certain historical fact from more than one perspective. For example, you might read about an event in "1 or 2 Samuel" and then read about it again in either "Kings" or "Chronicles." It may be written virtually the same, or depending about the time, it might be very different which definitely can be confusing. Now, with the chronological version of *the Bible*, you will see all versions one after another. And if David wrote one of his "Psalms" after an event, you will see it, and understand it, in context with what just happened. The same goes for *the New Testament*. There are some occasions where all four *of the Gospels (Matthew, Mark, Luke, and John)* wrote about an event in the life of Jesus Christ. They will be one after another, and some of them will be virtually the same, where in some situations, they may be quite different. I can understand why some people might think that there are many contradictions in *the Bible*. But there really aren't any. As a historian, I understand that it is all in the perspective of the individual that is writing their part. What one person sees, or writes, can be quite different depending upon what they saw, or heard, and how much or little they wanted to write down.

Take the *Book of Matthew*, for example. Some believe that it was written by the apostle Matthew. Others do not. The book was written anonymously. There is nowhere that it says it was written by someone that was actually with Jesus Christ. With the *Book of Luke*, on the other hand, we know that it was written by a man named Luke who was writing down facts that he had investigated so that he could tell them to a person named Theophilus. He says so in the opening of his book.

I love the simplicity of *the One Year Chronological Bible* and have read it, in full, more than once. I have read many parts of this book several times, especially the *New Testament* and in particular, *the Gospels of Matthew, Mark, Luke, and John*.

Chapter 22
Facing the Giant Teton Mountains

Facing the Giants left a lasting impression on me. In fact, when I went to return the DVD to the owner, he told me I could keep it. I was very grateful.

A few years later I took my first vacation to the Rocky Mountains during the winter. I made plans to fly to Jackson Hole, Wyoming, during the middle of December 2007. I wanted to take some pictures of the Grand Teton Mountains with snow on them. Every time that I had previously seen them, it was during the summer or early fall, and the taller mountains were snowcapped, but that was all. My great plan was to make my own Christmas cards, with the Tetons in the background and a decorated Christmas tree in the foreground. I was going to fly into Jackson, buy an artificial tree along with some ornaments, and then pull off the road and set it up, snap some pictures, and then make Christmas cards. I had even talked to Grand Teton National Park officials to get permission. Alas, as with so many other human endeavors, things did not go according to plan.

My flight was on a Friday, about two weeks before Christmas. I was scheduled to fly out of Nashville during the afternoon, catch a connecting flight in Dallas, and arrive in Jackson late that night. I would return on an early morning flight the following Monday. Now here is where God stepped in.

The plane that I was supposed to catch to Dallas was late. Finally, the captain made an announcement over the intercom. We couldn't go anywhere until engineers could inspect the damage to one of the wings and make sure it was safe to fly. I had been doing the math, in my head, and came to the realization that I was not going to make it to Dallas in time. So I left and talked to one of the airline officials. I asked if we could postpone my flight until the middle of January. He said yes. Although I didn't see it, at the time, here is where I see the hand of God working in the background.

Doing this allowed me to make a few changes. Instead of paying the super high prices of Jackson, Wyoming, I was able to purchase a relatively inexpensive Christmas tree at a local Walmart. I was able to pack it, along with cheap ornaments and even some Christmas lights in one large suitcase. My clothes and camera gear went into other baggage.

When my great day finally arrived, everything went a little better. There was no drama at the airport. I caught my plane on time, made the connecting flight in Dallas, with plenty of time to spare, and arrived in Jackson, Wyoming, at about 11:00 p.m. I picked up four-wheel drive rental car and made my way to the local Motel 6 where I was to spend the weekend.

There was plenty of snow. You don't know how happy I was to see the snow. In Nashville, we hardly ever see any significant amount of snow anymore. As my brother told me, it had started snowing in the Tetons, during the middle of November, that year, and had not stopped when I arrived there in the middle of January. Nor did it stop while I was there. Every morning I would wake up to a fresh two to three inches of snow on the windshield of my vehicle. And it would snow, off and on, all day, each day that I was there. In fact, I could not even see the Tetons, at all, on Saturday or Sunday due to the cloud cover. I could look out to the east and see the sun coming through the clouds, sometimes even quite sunny. But when I looked to the west, toward the Teton Mountains, all I could see was a wall of gray.

Don't get me wrong. It was a good trip. I even did some snowshoeing with a group out of the local visitor center in the park, and I took lots of "good" photos. But it was not the shots that I wanted to take of the Teton Mountains. Still, I was happy, and as I prayed to God, before I went to bed that last night, Sunday, January 20, 2008, I thanked the Lord for my blessings.

And I would have left it at that except for that movie *Facing the Giants*.

I remembered the scene where Coach Taylor's wife, devastated that once again, she was not pregnant, looked up at the sky and prayed. "I will still love you."

I know it's just a movie. And movies are made for a purpose, but that scene kept coming back. Her husband had once asked her if she would still love God even if they couldn't have children. That's why that scene resonated so much. *I will still love you!*

It's easy to be thankful when we get our way, but turn the tables, and we tend to less charitable. As I said before, that movie changed me, forever. It changed the way I think, the way I act, and most especially in the way that I see God. So as I said my prayers, that night, I said that I would still love God, even if I didn't get to see the mountains. And I was perfectly content to leave it at that.

But then I started thinking, why not? Why not ask God for a little extra? So I said, "Lord, if you don't mind I would really like to see the mountains. That's why I came out here. I want to take some pictures of the mountains. But it's okay if I don't. *I will still love you.*" And, then I went to bed.

The next morning, I noticed something different. At first, I could not put a finger on it, but there was something in the air. With a 3:30 p.m. flight on the horizon, I knew I needed to be at the airport by 1:30 p.m. With the obligatory 11:00 a.m. checkout time at the motel, I had several hours to kill. So I started watching some old reruns on television. Then, along about 10:00 a.m., it hit me what the difference was. *It is really bright outside.*

From the motel, I could not see the Teton Mountains. They're to the west of Jackson. I started to get my hopes up. Maybe I could see the mountains, today. My plans changed in an instant. Now, on a trip, I always pack everything up the night before, except for what few things I will need on my return flight home, so it was no time before I was rolling out the door.

And, lo and behold, what wondrous sight did I see when I finally came around that curve in the road—it was the most magnificent sight I had ever laid eyes on. That long row of 12,000- and 13,000-foot mountains that we know as the Grand Teton Mountains stretched for what seemed like an endless eternity, and they were covered in a carpet of freshly laid snow. Just like Coach Taylor felt when he came home after winning the state championship, and his wife told him that he was going to be a daddy, I was overcome with joy.

Every single pull-off that I came to saw me getting out of my vehicle and trudging through half a foot, or more, of snow and taking lots of pictures. I never even broke out my old tried-and-true 35-mm SLR camera. I had borrowed my friend's digital camera, and it took really fine pictures. I didn't want to go too far and possibly getting stuck so that I would miss my flight, so my last stop was at the Snake River Overlook, what I have always considered the best vantage point for seeing the Tetons. For these shots, I took out my tripod and set it up. Forget about trying to put up the Christmas tree, I knew there wouldn't be time. But it didn't matter. God had truly blessed me. I snapped shot after shot, from wide angle to zoom and back to wide angle again. Over and over, I merrily pressed that shutter button. And you know the best thing about a digital camera is you will know, right then, right there, if you got a good picture. No more waiting to turn in the film and waiting for it to be processed and then a few days later seeing if your pictures were any good.

And finally it was time to move on. It was time to head over to the airport, to fly home by way of Chicago, to return to the life I had left behind for such a short period. But I knew it would never be the same. The good Lord had blessed me.

I know to most people it might be just a few snapshots of some tall bumps in the land, but to me, I think it was a little more. Just like Stanley Praimnath asking the Lord to save him, on 9/11, or Genelle Guzman, reaching out to the God that she had turned her back on so many years before, I think, perhaps because I took that little extra time, and did ask for a little more, well what can I say? God does move in strange and mysterious ways.

Chapter 23
Church and the Golden Rule

I once heard a story told on the radio by the famed broadcaster, Paul Harvey.

There was a preacher at this church. He was young, mid to late twenties, so he resonated with the teenagers in the congregation. They liked him because he "got it." He had many different youth activities set up for them, and they liked this. One day, he introduced a guest speaker during his normal Sunday sermon. This man was a good bit older, in his fifties or sixties. After the brief obligatory introduction, the man started to speak. This man said that he wanted to tell them a story. This was much like the many parables that Jesus would recite to the masses that followed him. The story was about a father and his son, and his son's best friend. One day, they went fishing. The boys were about ten years old or so. I don't remember how it happened, but both boys wound up in the water, struggling and thrashing around begging for help. The father picked up a life preserver, but as he held it in his hand and started to throw it to his son, something stopped him. He thought about it, *I know my son is saved, but I do not know if his friend is.*

Struggling with his inner feelings, the man threw the life line to the other boy. He was able to rescue him and pull the young boy back to the boat. Alas; however, he was not able to save his son. The boy drowned. Crestfallen, and heartbroken like any parent would be over the loss of a child, the man still knew in his heart that he had done the right thing. Although grieving for his son, the man rejoiced in knowing that he had been able to save the other boy's life. And the man said that the other boy went on to have a good and productive life.

After the sermon was over, the older man was accosted by two teenage boys in the parking lot. One of them said, “Nice tale, Pops. You really had me going for a while, but there’s no way that any father is going to save another man’s son while his own boy drowns.”

The man stared at the boy for a long time and started to smile. Finally, he spoke.

“Well, I’m here to tell you this story was true. You see, I was the father. It was my son that drowned. And it was your preacher who was my son’s best friend.”

So let me ask you, what is “church?” Or what makes up a church? Is it because one is Christian, or Jewish, or Muslim, or even Buddhist? Is it because you belong to a certain ideology, such as Catholicism? Or is it the name in front of the word “church,” such as Baptist, Methodist, or Church of Christ? Can it just be a group of people, doing things for others, helping other people, because Jesus called upon them?

I saw an article by the Pew Research Center on the internet. It said that as of 2010, there were an estimated 1.6 billion Muslims in the world, which is roughly 23 percent of the population. There were 2.2 billion Christians in the world, roughly 31 percent of the population. That’s a lot of humanity. That’s over half of the world’s population. And they all started with one man—a Jew. A Hebrew named Abraham. I don’t know how many Jews there are in the world right now, but it is a small percentage of the population. And they were God’s chosen people.

Some 290 years after the great flood—Noah’s flood—a man named Terah had a son and named him Abram. God would later change that name to Abraham. And it was through Abraham that all of the earth’s descendants have come. Since Abraham did not defy God when he told him to sacrifice his son, this is what the Lord had to say:

> *Because you have done this and have not refused to give me your son, your only son, I will certainly bless you and make your descendants as numerous as the stars in the sky and the grains of sand on the*

> *seashore. Your descendants will take possession of their enemies' cities. Through your descendant all the nations of the earth will be blessed, because you have obeyed me.* (Genesis 22:16–18)

One day, God talked to Abraham. Now, Abraham wasn't like anyone else. You see, God really talked to him. The *Bible* doesn't say that Abraham had a dream or a vision. It's like two guys having a friendly chat. Why, you may ask? Remember that Abraham was a truly devoted servant to God. When God told Abraham to sacrifice his son to him, he didn't hesitate a second. And remember how long Abraham had waited to have his own son by his own wife, Sarah. Abraham was one hundred years old, when Isaac was born. For a century, the man that God said was going to populate the earth couldn't have a child by his wife. Yes, he had Ishmael, by his wife's servant, but not one of his own. And then after all the waiting, and all of the hope, God tells him to kill his son. Did Abraham say no? Did Abraham say why? Did Abraham plead with God? No, he just said okay. And he was ready to do what God told him to do. He had the firewood. He had the rope to tie his son; in fact he had already tied him. And he had the knife and was ready to plunge it into the young boy's body. At that moment, God knew he had a truly righteous man, one that loved him, so much, that he would take the life of his own son. We know how that ended. An angel stopped him and showed that a there was a ram in the bushes nearby.

One day, the Lord told Abraham that he was going to destroy Sodom and Gomorrah. He said, "Sodom and Gomorrah have many complaints against them, and their sin is serious."

Abraham replied, "Are you really going to sweep away the innocent with the guilty? What if there are fifty innocent people in the city? Won't you spare that place for the sake of the fifty innocent people who are in it?"

And the Lord said yes. He said that for the sake of fifty innocent people, he would spare the city.

Then Abraham asked if he would spare the city for the sake of forty-five or even only forty innocent people. Again, the Lord said yes. Continuing this line of conversation, Abraham asked if the Lord would spare the city for the sake of 3thirty, then twenty, and finally for the sake of just ten innocent people. And each time the Lord said that he would not destroy the city.

But as you well know, there were not even ten good people in Sodom and Gomorrah. Only Lot, Abraham's nephew, was found righteous in the eyes of the Lord. For the *Bible* says, "Before they had gone to bed, all of the young and old male citizens of Sodom surrounded the house. They called to Lot, 'Where are the men who came to stay with you tonight? Bring them out to us so that we can have sex with them'." That pretty much says it all; every man in Sodom was a homosexual. So God destroyed Sodom and Gomorrah.

Have you ever heard the joke about the Church of Christ? The joke is based on the supposition that they think that they are the only ones going to heaven.

It goes something like this. A guy lives a righteous life. Let's say he was a Baptist. He accepted Christ as his Savior, and so he receives eternal life upon the ending of this one. At the pearly gates of heaven, Saint Peter meets him, and as he shows him around, they see several groups of people walking and talking. Saint Peter points out one group as being Methodists, another as being Baptists. Over in a far corner he points out one group as being Pentecostals and another as being Catholics. But as they come around a corner, the newest member to heaven sees a big wall and asks about it. "Shhhhh!" Saint Peter says in a whisper, as he holds one finger over his mouth. "Those are the Church of Christ people. They like to think they are the only ones here."

And because I am a big University of Tennessee football fan, I must share this one.

Steve Spurrier, upon passing from this mortal plane, somehow made it upstairs. As he is being shown around heaven, by the archangel Gabriel, he sees these beautiful houses, some may even describe them as being mansions. The first one he sees is a very nice, modern-style house painted in the colors of Ohio State, Florida, and Utah. He recognizes it as belonging to Urban Meyer. Over to one side he sees the house of Joe Paterno, of Penn State fame, and in another area, he sees the garnet and gold house belonging to Bobby Bowden. Adorned in crimson and white are the twin mansions belonging to the great Bear Bryant, old style, and Nick Saban's more modern one. But then, as they pass up all the other houses, he sees a huge one that dwarfs all the others as far as size is. On a giant cloud, all of its own, this house/mansion looks like it must have at least one hundred rooms to everyone in all of the other houses. It is painted in the most beautiful shade of orange and white and has dozens of pennants flying from its rafters. And in the middle is an enormous pennant proclaiming the resident to be a member of the Big Orange Nation.

Spurrier is a little miffed about this, turns to Gabriel, and says, "Why does Phil Fulmer get such a large house?"

"Ah, but that's not Phil Fulmer's house," replied Gabriel, "That's God's house."

There was a movie called *Almost an Angel* with Paul Hogan of "Crocodile Dundee" fame as the main star. It was released in 1990. I guess you would call it a romantic comedy. There were a lot of humorous parts in the movie, but in general, the theme was salvation.

Terry Dean (Paul Hogan) was not a good man. In trouble with the law ever since he was nine years old, Terry had been released from prison, only to return to a life of crime. This time, however, he decided to advance to the big leagues. He became a bank robber. His different and unique style allowed him to pull off his first job and escape successfully. Shortly thereafter he saved a young boy who was about to be run over by an automobile. Dean himself, however, was struck by the vehicle. Taken to the hospital, he found himself in a hospital bed on a "cloud" when God walked up to him. As Terry remarked to a preacher, later on, God has an uncanny resemblance to Charlton Heston.

God/Charlton Heston looked down at a clipboard, which had Terry's life history on it. When Terry asks if he is God, his reply was, "Let me put it in a language you'll understand. You're on probation, Mr. Dean. Think of me as you're probation officer."

After going through a list of faults, God says, "Your instinct has earned you another chance. Probation, Mr. Dean."

"How does that work?" Terry asked.

"You will go back and dedicate yourself to helping others. Giving, not taking. You will become an angel of mercy."

"Probationary," God adds, "Strictly on a trial basis."

However, when Terry asks God if this happens all the time, God replies, "From time to time, but these are difficult times, Mr. Dean. In this century, you are the first scumbag we've sent back."

The underlying core value of this film, however, goes back to "The Golden Rule" that many if not all of us were taught as a child. "Do unto others as you would have others do unto you." Early in his ministry, Jesus went up on a mountain and began teaching his disciples. A very large crowd had also followed them. In what would become know, as "The Sermon on the Mount," we learned many of the ideas and thoughts that would govern our lives. It was from this, that we learned the Beatitudes, blessed are the meek, for they will inherit the earth; blessed are the peacemakers, for they will be called children of God, and "The Lord's Prayer."

God wants us to ask him for help. He will help us.

Jesus said, "God answers all prayers."

Sometimes the answer is yes.

Sometimes it is maybe.

And sometimes, it is no.

God hears all prayers. God answers all prayers. Sometimes it's "maybe." Sometimes it's "not now." Sometimes it's "let me think about it," *and* sometimes it's just plain "*no*!"

But you know what. That's alright. I have learned to accept what God gives me and to be thankful for it, because God has the *big picture*. He has a plan, and he has a reason for everything. I'd like to be rich and never have to worry about how I'm going to pay my bills again, but if I don't, that's okay. The greatest treasure that any of us can receive is God's blessing and eternity in his heavenly paradise. I don't know about you, but on judgment day, I do not want to hear Jesus say, "I don't know you."

Some folks think God gave him a brain to solve their own problems.

Other folks pray only when they hear a siren, for whoever needs there help.

Remember Abraham asked God to spare Sodom and Gomorrah for the sake of fifty, then forty-five, then forty, and finally all the way down to ten righteous men. But he couldn't find ten righteous men, so they were destroyed.

If I ever had the money, I would like to build a church. That is where I would go on Sundays. It would not be fancy. It wouldn't even be very large, just a place to worship the Lord.

My friend, Steve, in describing his church, said, "Oh, it's not very large."

"Well, how many people go there, Steve?" I asked.

"Not that many, no more than five hundred or so."

"Uh, Steve, when I was talking about small, I was thinking more like fifty to one hundred people."

My church would not be big. It wouldn't be about showing off our clothes, or our cars, or where we were going to eat dinner after church. We wouldn't even have a big-time preacher.

It would just be a place where people could go and learn more about God and Jesus Christ. If someone had a certain passage from the Bible that meant a lot to them, they could stand up and read it, and then we could talk about what it meant and how it helped us. Each Sunday might see a different person stand up or not.

I have learned a lot since I started studying the Bible. This was the legacy that I received from watching *Facing the Giants.* I know that I do not know it all. That is why I want to learn. More about God, more about Jesus, and more about what it means to me.

The name will be "church." Not the First Baptist or First Methodist Church or even the First Church, just church.

God is God, and that says it all.

Acknowledgments

I (Steve) want to acknowledge the people who have inspired me throughout my life and during the writing of this book.

Matthew and Elizabeth, my son and daughter. A father couldn't ask for two better kids. I am so proud of you both. We have shared so many great experiences. Myong Lee, my second wife, and the love of my life. She is indeed beautiful both inside and out. She has been a great supporter of my various projects over the years.

Robert E. Lee Schmittou, Bobby, one of my best friends for over forty years. He's an excellent writer (mainly the Old West fiction). Bobby is a contributing author to this book. Not only would he give you the shirt off his back, he is also a great chef. Thank you, Bobby.

Andy Pinkerton, one of my best friends of almost fifty years. We first met at the orphan's home. He is a self-made man, building a stone mason business from scratch. You inspire me, Andy. Mark Curtis, a fine CPA and member of the Church of Christ. Mark has been a friend for over thirty years. He has been so kind as to work on several of my projects on a "sweat equity" basis. Buck Dozier, a partner with me in business. He served for years as Nashville's Fire Chief, and he ran for mayor of Nashville. I have a great deal of respect for him. He is an elder in the Madison Church of Christ.

Lucius Burch III, the best venture capitalist in the entire southeastern United States. He was a true role model for me over the years, always taking my requested meetings and returning my calls.

Hugh Tharpe, Manager of Franklin Cigar, in Franklin, Tennessee, a favorite hangout of mine. Hugh is a very good friend. Even knowing him only a few years, he's the kind of guy you feel like you've known all your life. He has been a constant supporter of mine over the last couple of years, while writing this book. Like myself,

Hugh is a big supporter of University of Kentucky basketball, having actually worked as a detective at U.K. before.

James Everett Meeks, another great friend for almost forty years. He is a computer genius who was invaluable with this book. Joel Kneedler, my editor on this book. I could not have finished this book without him. He has become a very good friend in a short time.

Thank you all.

Song List

Bobby and I wanted to share a brief song list with you. Here are a few songs that inspire us, and we think they will inspire you too. Most of these can be found on YouTube or iTunes with a simple search online. We hope you enjoy them.

- ✓ "Amazing Grace" by Wintley Phipps at Carnegie Hall
- ✓ "Elvis American Trilogy"
- ✓ "God Bless the USA and Armed Forces Medley"
- ✓ "How Great Thou Art" by Carrie Underwood and Vince Gill
- ✓ "It Is Well with My Soul" by Guy Penrod and David Phelps
- ✓ "Oh, Holy Night" by Celine Dion
- ✓ "Revelation Song" by Philips Craig and Dean
- ✓ "Shout to the Lord" and "Agnus Dei" by Darlene Zschech
- ✓ "The Americans" by Gordon Sinclair
- ✓ "I Can Only Imagine" – Mercy Me

About the Author

Within *Reasons to Rejoice*, Steve Rydell gives the accounts of the highlights of his life, a rather exciting ride it has been. Having grown up in an orphanage from the time he was eight, life for Steve has not always been easy. Throughout his life he has had experiences involving miracles and angels. The hand of God literally saves his life several times. Professionally, Steve has worked in the following fields: as a certified public accountant (CPA), a business broker, a commercial realtor, and several others, even as a bouncer at a nightclub. Mr. Rydell actually traces his life as it ties to the religion he was brought up in—the Church of Christ.

CPSIA information can be obtained
at www.ICGtesting.com
Printed in the USA
LVHW050313060619
620307LV00002B/2/P